CAD/CAM: FEATURES, APPLICATIONS AND MANAGEMENT

Peter F. Jones
CAD/CAM Specialist

First published 1992 by
THE MACMILLAN PRESS LTD
Houndmills, Basingstoke, Hampshire RG21 2XS
and London
Companies and representatives
throughout the world

ISBN 0–333–48531–9 hardcover
ISBN 0–333–48532–7 paperback

A catalogue record for this book is available
from the British Library.

Printed in Hong Kong

Reprinted 1993

To Lelia for doing without while I was writing this, and CAD managers everywhere who have to make it all work.

Table of Contents

List of illustrations

List of tables

Preface

Since this book is about obtaining value from computer-aided design (CAD) it should be useful to student or established professional alike. I have had several kinds of reader in mind: those who want to start their career with a good understanding of new techniques, someone who may be drawn to the prospect of managing a CAD system, the experienced engineer who has been asked to procure or manage a new system, and the senior manager or director considering its introduction. All have an interest in the profitable use of CAD.

CAD has matured. The early exaggerated sales claims have given way, not surprisingly, to a mixture of successful applications, expensive failures and reasonable run-of-the-mill usage. To many companies it is now a normal tool to be used by designers and it is as such that it should be included in engineering courses. I have therefore treated CAD as a tool. When you use a new tool you mainly want to know what it will and will not do, and to this end it is useful to know something of how it does the job. (A knowledge that a spanner is transmitting torque provides insight on what the flats of the nut are experiencing, for example.) You also need to know not just how to use it but how to use it efficiently, and to understand the different variants of the tool and to judge its quality. (Ring spanner, open-ended spanner or box spanner and is it strong enough?). Finally, all tools in professional use need proper management to prevent them changing from valuable aids into expensive liabilities due to errors, accidents or wear. (A worn spanner can damage nuts.)

The particular outlook adopted here has led to a greater range of topics than many previous books on the subject. For instance, data security procedures and some important and little-known material on long-term data storage has been included on account of the exceptional longevity of engineering designs. The management politics of procurement is also discussed since many technical people tend to be unaware of such issues.

The book starts with an account of the way computers do their job principally for the benefit of older engineers who have not had the opportunity of understanding the basics of digital computers. Some readers will wish to omit this chapter.

The second chapter discusses some of the terms that are bandied about and the value (and pitfalls!) of using CAD.

Following these introductory chapters we describe and discuss the various items of equipment to be found in a CAD installation.

Next follows a full presentation of the many functions offered by CAD software. An effort has been made to indicate the contribution a function makes to the successful exploitation of CAD to allow those selecting a system to make informed judgements on what is being offered by the various vendors they are considering. The section includes a chapter on the specialised electronic CAD systems. As transferability of CAD models between companies is going to become increasingly important, three chapters on standards have been provided. Finally, the all-important but often neglected facilities to aid the management of the system are discussed and the section concludes with another vital but neglected topic: software quality, reliability and robustness.

All this material is designed to provide a good technical background to evaluating and procuring a system but procurement involves management issues and these are covered in the chapters which follow.

Finally, having procured a system, it has to be managed well to achieve results, and so the book concludes with chapters on this important aspect.

The world abounds with ill-defined buzz-phrases and acronyms amongst which are CAD and CAM (Computer-aided manufacturing). It is therefore important to say at the outset how I will be interpreting them for the purposes of this book. I am assuming that the reader will be concerned with a commercially available "CAD/CAM" system intended for use by designers. Such a system often provides facilities for the geometry defined by the designer to be converted into control instructions for numerically controlled machine tools. This process is described and its relationship with manufacturing procedures is discussed but the detail of part programming machine tools is not covered on the grounds that it is usually the concern of manufacturing specialists. Designers need to know how their CAD models may be used to supply data to Manufacturing but they will not normally be involved in actually programming the machine tools.

Peter F Jones
January 1992

Acknowledgements

When attempting to cover a wide area such as CAD/CAM, you can either farm out chapters to other authors or attempt to do the whole job yourself. If you do the latter there will always be some topics where you need to consult colleagues for their knowledge and experience. I am therefore grateful to those at Cranfield Institute of Technology and elsewhere who have allowed me to pick their brains. In addition, I am grateful to those who have provided material for illustrations and those who have read parts of the draft to check the level of presentation.

The CIM Institute Mr Peter Stokes, the Chief Executive, gave permission to use the institute's facilities in preparing the book, and read and commented on the draft.: The IBM Document Composition Facility, with which the book was typeset and the IBM Catia CAD system and Display-Graphics personal computer graphics program contributed to the later stages of the book which was produced, with the exception of certain pictures of equipment drawn by a professional illustrator and two pictures copied from manuals, entirely electonically. The text was written using the Generalised Markup Language. This allowed full transferability between the personal computer, where the text was originated and the mainframe, where it was formatted to the typographical specification of Macmillan.

Dr Stephen Evans made a valuable contribution to the chapter on numerical control from his experience of setting up numerical control systems in industry.

Mr Raymond Goult provided a valuable introduction to IGES and STEP from his work on the international standardisation committees.

Ms Elizabeth McLellan contributed much useful information on numerical control programming from her experience in teaching and carrying out numerical control programming and in writing post-processors.

Mr Frank Ainscow contributed a very useful account of the problems in designing and manufacturing colour cathode ray tube displays while on secondment to Cranfield Institute of Technology from IBM.

Mr John Heath contributed useful information on the characteristics of magnetic media and introduced me to the valuable US Bureau of Standards report on the topic. I am also grateful to all those who have provided discussion and comment on various portions of the book including Dr Peter Deasley, Dr Victor Newman, Dr David Kirk, Mr Paul King, Mr Peter White and Mr Bob Almond. The photographs of equipment were taken at the CREATE design consultancy at Cranfield. Mr Harry Wall assisted in setting up the printing software while at the IBM office in Warwick.

Chapter 1 An introduction to computers

If you are reasonably comfortable and familiar with computers then skip this chapter and start with "CAD/CAM and its value" on page 13. This general introduction to computers is included for the benefit of those who find the arrival of computer technology in their immediate working environment a little disturbing - and not without good reason, for CAD/CAM is computer technology with a vengeance. It is ironic that engineers, who have been the agents of the technological revolution in society, should now undergo a sharper change in their working methods than anybody else has ever experienced!

In the film "2001" there was a computer called HAL controlling the spaceship. Its circuit boards were pulled out one by one. It was a talking computer and its behaviour gradually became more and more infantile until it ended up reciting nursery rhymes. Although computers might have human behaviour like HAL one day, nothing could be further from the real nature of computers as they are today, even after 30 years of development.

The fact is that computers are machines. Just as mills, lathes and moulders are machines for converting metal or plastics from one form into another, computers are machines for converting information from one form into another. Both are completely deterministic in their operation, remorselessly following the sequence of movements or operations defined for them by their designers.

However, there are certain characteristics of computers which make them different from mechanical devices. Firstly, a computer is able to follow a very long and complicated sequence of operations. Secondly, it possesses the ability to take different paths through the sequence depending on the values of particular items in its data. This allows computers to reproduce in a deterministic way the power of humans to make decisions, and computers are therefore capable of very complex patterns of behaviour. Thirdly, the sequence of operations is held in an electronic *store* (also known as the *main memory*) which means that the sequence can be changed quickly and easily. The sequence for performing one task can be rapidly replaced by that for performing another. A computer can thus be switched from one job to another very easily. Finally, the operations are performed extremely rapidly. As many as a million operations can take place in a time frame which a human would regard as instantaneous.

A computer is thus a device, restricted to performing transformations on information in a solely deterministic way, which is nevertheless extremely fast, flexible and versatile.

Two distinct parts of a computer can be identified from the previous account, one tangible and the other intangible. The intangible part is the sequence of operations. It is intangible because it consists only of electronic signals in the memory but it is essential because it determines what the computer does. The tangible part is the electronic circuitry which carries out each operation in turn in the sequence and which provides the memory. The tangible part is called the *hardware* while the intangible sequence of operations is called either the *program* or the *software*. The term hardware is also used for the other physical elements of the computer such as printers, terminals, cases, cables etc.

THE HARDWARE

A large part of the hardware is built up of millions of a basic electronic circuit. This circuit has the simple property of being either fully on or fully off (rather like a push button or light switch). Every item of information is represented by a row of these circuits holding a particular pattern of "ons" and "offs" and the activity of the computer consists in turning these circuits on and off to represent the changing state of the information. A row of "ons" and "offs" is known as a *binary code* and is written down as a row of 0s and 1s. Each 0 or 1 is known as a *bit* (short for binary digit). Each basic electronic circuit is therefore used to represent one bit. Table 1.1 shows how the numbers 0 to 10 are represented in binary codes. If you look at the sequence of codes it is a bit like trying to count without being allowed to use the numbers 2 to 9! You can also see that it takes four basic circuits in a row to hold just one digit. However, with thousands of micro-circuits on small silicon chips the large numbers of devices required present no serious problems.

These are the codes used for numbers, which of course play a large part in computers. Other tables of codes are used for other kinds of information. A particularly important kind is text for which an internationally standardised code, known as the *ASCII* code, is used. ASCII codes are 8 bits long with the result that 8-bit binary codes are extremely common and have been given the name of *byte*.

There are two varieties of the basic electronic circuit: one which stays on or off once it has been set, like a light switch, and one which responds to the particular signals it is receiving at the time (rather like a bell-push). The latter type is called *logic*. Logic circuits perform the actual transformations on the information, converting a binary code applied to their

Table 1.1 Binary codes

Number	Code
0	0000
1	0001
2	0010
3	0011
4	0100
5	0101
6	0110
7	0111
8	1000
9	1001
10	1010

inputs into a different code at their outputs according to the particular transformation each is designed to do. The former type which stays set is called a *register*. Although the term can be applied to a single bit it is usually applied to a row of bits holding a binary code of a particular length.

When you consider that all of the program and data has to be stored electronically for it to be used by the computer you can see that by far the greatest part of a computer is given over to memory. Because of the large amount required, several different kinds of device are used for storage in order to economise by striking a balance between speed and expense. While an arithmetic operation is actually in progress the two numbers involved are held in just a few very fast registers capable of operating in a fraction of a millionth of a second. On the other hand, the program and data uses hundreds of thousands of registers which can be slower and therefore less expensive: operating in about a millionth of a second. This is the main memory. The word memory is also used on its own to refer to the main memory and this will be the usage of the word in the remainder of the book. Figure 18.3 on page 171 shows a main memory circuit board with a capacity of 2 megabytes (2 million bytes or "2 Mbyte"). Naturally, the length (number of bits) in each register of the main memory is the same for a particular computer and is known as the *word length*. Since binary codes are being transferred from the main memory through the logic and back to main memory the same word length is used throughout. The word length thus characterises the computer as, for example, 16-bit or 32-bit etc.

Programs and data which are waiting to be used in the next job need to be held in electronic storage so that they can be transferred to the main memory when required, but a slower speed than that of the main memory is sufficient. A larger capacity is required so that many programs can be stored. A *disk* memory employing magnetic recording which can be accessed in around a hundredth of a second is used for this purpose. Disk memories come in a wide range of sizes and types from *"floppy" diskettes* (see Figure 7.5 on page 69), through *hard* disks as used in a Personal Computer to large *Winchester* disks on a large data processing installation. The capacities range from the equivalent of about a third of a million characters of text to several thousand million characters. Figure 7.7 on page 71 shows a hard disk.

Finally, for the slowest access speed, largest capacity and lowest cost of storage there is the magnetic tape unit (see Figure 7.6 on page 70). It uses a reel of tape costing about $15 which is capable of holding the equivalent of a hundred million or so characters of text and taking about 15 minutes to read.

The rest of the hardware will comprise printers for printing out the results, ancillary units such as power supplies and cooling fans, and possibly communications devices for allowing access to the computer over telephone lines.

THE SOFTWARE

The part of the computer which determines the job it performs is the sequence of operations held in the main memory. A single sequence is called a program or routine and a collection of programs is referred to as software. A single operation in a computer is called an *instruction.* It will be a binary code specifying what transformation is to be performed on what piece of information. A typical instruction might take two numbers from the data and add them together, leaving the result in a register for further use. Another instruction might stow the result away in main memory again or perhaps perform an operation on a single number such as negating it. For non-mathematical work there are instructions which slide the binary code patterns from side to side, for example.

Because each individual instruction performs only a very basic operation, many thousands are required to do anything useful. Designing and writing a program involves breaking down the task into a long complex sequence of basic operations, each of which has to be accurately specified. To make this task manageable, a powerful technique has been devised.

Since instructions are binary codes they can be treated as information to be processed just like any other information, which means they can be generated by another program. You can make programs which generate further programs! This opens up the possibility of having a program which reads a mathematical formula typed on the keyboard and translates it into the sequence of instructions which do the calculation. In fact, one usually wants the computer to do more than evaluate a formula, such as make decisions, so to describe the full range of what might be required in a form which can be easily translated, a special language is devised.

Programs which perform this translation are called *compilers* and the languages which have been invented for specifying what the computer is to do are called *high-level languages*. A high-level language is designed to be as close as possible to the way humans express a problem while at the same time being sufficiently regular and rigorous for the compiler to translate without error and ambiguity. Examples of such languages are FORTRAN, designed for scientific and engineering calculations, BASIC, a simplified version of FORTRAN, and COBOL for commercial data processing. A portion of a FORTRAN program is shown in Figure 1.1.

The program as typed by the human programmer is usually called the *source code* and the binary code coming out of the compiler is called the *object code*. Usually, the source code is translated in sections producing object code in separate pieces which are then linked in an operation called variously *loading*, *link editing* or *link loading*. When ready to go into the memory to be run the program consists of the computer instructions and in this form is called *machine code*.

Most compilers read the whole high-level language program before doing the translation but there is a special type of compiler which reads just one line before translating it and executing it. This is called an *interpreter* or *incremental compiler*. Its best known use is in BASIC but it is very often used to implement interactive and easily used languages.

The size and power of computers is such that they can handle really large programs - some of them are so large and complex that they create problems for the programmer in understanding fully what he is producing. This is the case with CAD software, so it is worth looking at some of the methods used to make programming more manageable and at what makes software break down.

There are many occasions in daily life when we treat a sequence of actions as just a single operation because it occurs so often that we do it automatically. For example, making a telephone call consists of a sequence of looking up the number, dialling the number, asking for the person when the phone is answered, dealing with engaged or wrong number situations etc. Each call is different but the basic sequence is the same. Once we have learnt it we can perform it without thinking. Also, regarding it as a single operation simplifies the planning of our work.

The computer provides a similar means of simplifying a program using what is known as a *subroutine*. For example, if there was a program which calculated the area of a number of various circles in the course of its operation, the sequence of instructions for doing that calculation would be stored once only in the memory. Each time the calculation is required the computer breaks off from the main sequence, noting where it had got

```
C      CALCULATE THE ELEMENT ORIENTATION
C      This calculation is performed by examining the signs of the
C      values in the transformation matrix.
C
       INVRAD=3.141592/180
       DEGCON=0.0174533
       PI=3.1415927
C
       IF (TRA(1).GE.0.0.AND.TRA(2).GE.0.0.AND.TRA(4).LE.0.0.AND.
     1TRA(5).GE.0.0) THEN
             ALPHA=ELEORI(I)
       ELSEIF (TRA(1).LT.0.0.AND.TRA(2).GT.0.0.AND.TRA(4).LT.0.0.AND.
     1TRA(5).LT.0.0) THEN
             ALPHA=ELEORI(I)
       ELSEIF (TRA(1).LT.0.0.AND.TRA(2).LT.0.0.AND.TRA(4).GT.0.0.AND.
     1TRA(5).LT.0.0) THEN
             ALPHA=(2*PI)-ELEORI(I)
       ELSEIF (TRA(1).GT.0.0.AND.TRA(2).LT.0.0.AND.TRA(4).GT.0.0.AND.
     1TRA(5).GT.0.0) THEN
             ALPHA=(2*PI)-ELEORI(I)
       ELSE
       ENDIF
C
       ANGLE(I)=ALPHA/INVRAD
C
          DO 5 J=1,3
             PLAPOS(I,J)=TRA(J+9)
  5       CONTINUE
C
       WRITE(6,920) ELEJEL(I),ELETYP(I),
       WRITE(6,930) PLAPOS(I,1),PLAPOS(I,2),PLAPOS(I,3)
C      Writes out the x,y,z coordinates for each element placed
       WRITE(6,950) ANGLE(I)
C      Writes out the orientation of each element in radians
 920   FORMAT(5X,I4,5X,A16)
 930   FORMAT(5X,F10.3,5X,F10.3,5X,F10.3)
C      XYZ coordinates for the elements placed on the table
 940   FORMAT(5X,F10.3,5X,F10.3)
 950   FORMAT(5X,F10.3)
```

Figure 1.1 Part of a FORTRAN program

to, and jumps to the beginning of the area calculation sequence. At the end of calculating the area the computer looks up where it was in the main sequence and resumes at that point. The area calculation sequence is called a subroutine and the act of jumping to it a subroutine call.

The value of this technique is not just that it saves repeating the same set of instructions but that the area calculation, for instance, can be written separately as a self-contained task, thus allowing the program to be broken down into manageable units. The technique also makes programming more manageable in another way.

In programming calculations you need to give a name to each variable. In a large program this can involve many names and the danger of using the same name twice. The use of subroutines solves the problem by arranging that any names used by the subroutine have no meaning outside. The temporary data used by the subroutine cannot be accessed by the main program. Such data is called *local data.* Special provision is made for passing over the names of variables the subroutine is to work on and for passing the results back.

A computer makes a decision by taking an alternative path through the sequence of operations according to the value of a variable. In a typical program, many decisions are taken so that there are many alternative paths. Instead of just one clearly defined path through the long sequence of instructions there can be hundreds of alternative paths and thousands of combinations of those alternative paths! A program can be just like a maze and it is possible for a path to lead back exactly to the place where it started, in which case the program will just go round and round in circles and never stop! This characteristic of programs which makes them like mazes is the biggest source of problems in programming.

Modern high-level languages have certain features to alleviate the situation. Firstly, they make alternative paths join up again at a given place, so the programmer knows where the main sequence will be resumed whichever alternative is taken. Secondly, where a programmer deliberately wants to go round and round in order to repeat a calculation he is given a means of defining which sequence is to be repeated and how many times. These facilities prevent the program from becoming a maze and allow one to follow through the various alternatives the computer might take. As they work by clearly identifying the beginning and ending of portions of the program, languages using these features are called *block structured.*

Unfortunately, there is another problem which is not so easy to solve. It shows itself in a paradox. Although there is absolutely no way in which software can wear out like mechanical devices, and a computer is absolutely deterministic in operation, software nevertheless breaks down randomly like a mechanical device! The reason for this is again the many alternative paths and it can be explained as follows.

Testing is a major part of software development. You cannot be sure of the correctness of any sequence of instructions until the computer has run through them and the result it obtains has been checked. But only one of the many alternative paths can be followed each time the program is run. Although the programmer runs tests which make the program take as many of the possibilities as he can, there will always be some paths unexplored. Of these, one or two will have mistakes in them. After testing, the program goes out to the users who run it over and over again. Eventually, a user puts in a particular set of data which takes the computer down an untested path in which there lurks a mistake and suddenly the program fails. We shall be returning to this matter later when we discuss the reliability of software.

CAD is a very complex application of computers to be placed in the hands of those who are not trained computer professionals. In this chapter we have given a simple picture of how computers work and the particular strengths and problems that result from the way they work so that informed decisions can be made to obtain the best use of the technique.

EXERCISE

Describe the ways in which computers are similar to and the ways in which they differ from human thought processes.

Chapter 2 CAD/CAM and its value

CAD/CAM SYSTEMS

CAD stands for Computer-Aided Design. We can think of a number of ways in which computers could aid designers. These are:

1. Performing design calculations

2. Producing and managing parts lists

3. Storing and retrieving design information

4. Producing drawings

5. Producing programs for numerically controlled (NC) machine tools

Design calculations and parts list processing were the first applications. They were fairly obvious things to use computers for since they involved processing well organised data in well defined ways and the data was just text and numbers.

The use of computers for information storage and retrieval is sadly neglected. Although designers spend a significant amount of time searching for data on standards, test results or the characteristics of components, computer-assisted information retrieval has still not been applied to any great degree. This may be because the data is not very well structured and the procedures not well defined. Nevertheless, the techniques for storing and retrieving large amounts of unstructured information are well known to librarians and information officers who have developed advanced computer software for the purpose. For some strange reason this technology and experience has yet to be applied to the smaller world of the design office.

The production of drawings, designs and NC programs using computer aids, which is the subject of this book, seems to have become the activity that many people mean by the term CAD. This is probably not so much on account of its usefulness as of its spectacular nature and the misconception that because drawings are what come out of a design department the designers spend all their time drawing. The following similar abbreviations are currently in use:

CAD Computer-aided design

CAE Computer-aided engineering

CAM Computer-aided manufacturing

As can be seen from their full names, they refer to the use of computers to assist in the three activities respectively. In this book we shall restrict ourselves to the facilities which are currently provided on what are usually called CAD/CAM systems. Without wasting space on what might be the precise definition of the term "CAD/CAM" we can characterise the equipment sold under this description as really geometric data processing systems for engineering. They rarely handle non-geometric data except in association with geometric data, they do not handle manufacturing data except that which is directly derived from the geometric data, they are used almost exclusively by the engineering department and they usually reside there.

ELECTRONIC PENCIL OR PRODUCT MODELLER?

CAD/CAM has been sold principally as a way of increasing "productivity" in the Design Department. This emphasis has unfortunately led to a misconception of the nature of CAD and obscured an important benefit with the result that many users do not take full advantage of it. On the surface of it CAD looks very much like a super electronic pencil - a kind of graphical word-processor with all sorts of ways of producing lines on paper very quickly. But below the surface something much more significant and very different from making lines on paper is taking place. The computer program supporting the graphical effects is doing very precise calculations and storing the location of each point and the parameters of each line to a degree of precision only possible with a computer. Thus, if the designer draws two parallel lines 20 mm apart, they are recorded by the computer program in its data area (i.e. the CAD drawing) as 20.0000 mm apart and with an angle of 0.00000°. This is different to a paper drawing where the lines will be 20.1 mm apart because the designer cannot position his pencil more precisely than 0.1 mm.

The accuracy to which the data is recorded is actually higher than can be achieved in the workshop for most CAD systems. The result of all this is that the data represents the real world far better than the scaled drawing on paper attempted to, and in fact better than the manufacturing process will eventually achieve. Far from being an alternative to a piece of paper, a CAD drawing is a model of the actual product being designed. Because of this the designer can analyse it for clearances, volumes and surface areas etc. The drawing becomes what the paper drawing attempted to be: a truly precise definition of the product.

Now although CAD is able to be precise it will not be precise if the designer still treats the screen as a sheet of paper. If two hole centres are supposed to coincide the designer can either locate one point over another by positioning his cursor by eye or he can ask the CAD program to make the two points coincide. The results in either case may look the same but only one is a precise model which will give correct answers for analysis.

THE BENEFITS OF CAD/CAM

CAD gives benefits in three areas:

1. Quality of design
2. Design labour
3. Lead time

Better quality is obtained in two ways. Firstly, the overall accuracy of the design is improved so that less errors are discovered when it is implemented. Secondly, a better product can be designed. Labour can be reduced in two ways also: firstly, the time taken to generate or modify lines on paper is reduced and secondly, the labour required in other parts of the company to extract the information needed from the design can be reduced. Finally, the lead time is reduced because it takes less time to produce drawings.

Going into more detail the overall accuracy of the design is better because a CAD "drawing" is actually a precise mathematical model of the real object being designed. A paper drawing was to some extent a model of the real object because it was drawn to scale but the accuracy of the representation was severely limited by the quality and thickness of the pencil lines. In a CAD drawing, every point has a very precise pair of coordinates so that the distances between points and the lengths of lines etc. can be calculated exactly. Two advantages follow. Firstly, the designer can get the software to give him precise information about his design at any time. In particular he can find out clearances which would be invisible on a paper drawing. This helps in developing his design. Secondly, the "drawing" becomes a complete and unambiguous statement of the design on its own. This is particularly apparent when dimensioning. The computer itself writes the dimension over a line by calculating the actual length between the points at the ends. In a paper design the length has to be remembered or calculated from other design data. The dimension only represents the distance the designer intended if he takes care.

But it is not only in the geometric data that a CAD drawing can become a complete definition of the design. The large size of present-day computer storage means that the drawing can hold much more data than a sheet of paper. As much qualitative, textual non-geometric data as is needed can be held along with the geometric data. The processing ability of the software means that, although a very large amount of data is present, only that required for the particular purpose in hand is displayed at any one time, thus avoiding the confusion you would get with a large sheet of paper cluttered up with printing.

So far we have only considered the quality of the drawing as a definition of the design. But the thing itself can be designed better. The precision of the design allows accurate design calculations to be made by the computer and, because they are done faster with less human effort, more alternative cases can be examined in the time available. The speed with which modified versions of a design can be produced also allows more options to be examined in the time available. Besides this there are certain things which are almost impossible to define on paper. Complex doubly curved surfaces present great difficulties simply because of the limitations of representing three-dimensional objects on paper. CAD enables them to be defined precisely and visualised easily as can be seen in Figure 4.1 on page 32. Another example is provided by integrated circuits where the number of components on the chip would be just too big for a sheet of paper.

Turning now to the labour and time-saving properties of CAD the work in producing a drawing is reduced simply because an engineering drawing is composed of many almost identical simple outlines. With CAD, a copy of an outline can be used again in another position. New outlines can be produced by transforming previously drawn outlines with magnifications and rotations. In the same way the work in modifying an existing design is considerably reduced by using these transformations.

But it is not just in the Design Department that labour can be saved. CAD/CAM provides potential labour savings in other parts of the company as well. This comes about because of a combination of two factors. Firstly, a CAD design is readable by other computers and secondly, it can be a complete and precise definition of the product in a way that a drawing cannot be. Complete because the CAD "drawing" can hold as much geometric and non-geometric data as is needed and precise because the geometric data is held as numerical values. The computers used in other departments can therefore read the design, extract the relevant data and then process it into the form they require, thus saving clerical effort and avoiding clerical error.

We have outlined three areas of benefit: quality, labour reduction and lead time reduction. Which of these three areas of benefit is the most profitable? The factor most often used for the economic justification of CAD is that of labour reduction or productivity. This is probably because it requires expensive capital equipment in a department which previously used quite inexpensive equipment - drafting machines and boards. Furthermore, the equipment is revolutionary, dramatically changing the way designers work. The expense and the lack of precedent mean that a thoroughly credible case for the purchase has to be made in advance, leading to the need to project quantifiable savings.

But although the productivity case is impressive because numbers can be quoted and precise-looking cost-benefit calculations done, it is not a good one. To start with, the productivity is only obtained for the time actually spent putting lines down on paper. Observation of designers actually at work reveals that only a part of the designers' time is spent in this activity. They spend significant amounts of time on other things such as looking up specifications and standards, doing calculations and conferring with colleagues. For example, a recent study on one firm showed the distribution of time in a Design Department outlined in Table 2.1. This means that any productivity gain which might be obtained from the CAD/CAM system is diluted by the other activities: in the example given by around 70%. The productivity case can be further weakened by the rather low gains obtained in certain areas. For instance, in original mechanical design it is normally less than 2:1 so that one person combined with a CAD/CAM system produces a little less than that produced by two people. To get a net financial gain the equipment would have to cost significantly less than one person to run. The productivity gain would be marginal but with care, careful choice and efficient use, one could get the equipment at least to pay for itself. However, there are certain activities where clear gains can be obtained and the productivity case is strong. These are parameterised drawing, where the designer supplies a few parameters and the system produces a drawing or a set of drawings in return, and the well established area of electrical draughting and printed circuit board design. The conclusion to all this is that one has got to work hard (or cook the figures!) to present a good productivity case.

Though not quantifiable, a far better case comes from considering the effect on quality. Here, substantial gains are possible. Taking the improvement in accuracy, a design is issued, and arising from it materials are purchased, metal is cut and labour is expended. The discovery and correction of errors becomes increasingly expensive as time goes on. There is thus a high value on avoiding errors in the design. The errors need not

Table 2.1 Distribution of time in a Design Department

Activity	% of day spent	
	Designer	**Draughtsman**
Geometric manipulation	27	37
Searching for drawings	9	30
Calculations	30	3
Looking for written data	13	14
Internal meetings	5	3
Text creation	7	5
Personal	9	8

be just slips in dimensioning which are usually picked up in the checking process but clashes between components due to an inadequate appreciation of spatial relationships or due to piece parts being designed with inadequate reference to the original layout design.

Then there is the opportunity to design a better product. In commercial competition, a better product can radically change the fortunes of the whole company simply by tipping the competitive balance, whether it is better because of its performance or because it is cheaper to manufacture.

A good case can be made on lead time reduction for companies in the capital goods business where contracts for specially designed equipment are obtained by competitive tender. Presentation, timeliness, responsiveness to customer requirements and a competitive time scale are all factors made available by CAD/CAM and they can confer large benefits by tipping the balance.

To sum up, the biggest benefits are to be obtained from the various quality improvements possible although these are not quantifiable. For some operations, significant labour savings are possible and there will be certain companies where the lead time reduction is the main factor.

THE DISADVANTAGES OF CAD/CAM

The combination of hard selling and novelty has resulted in a certain amount of hype being applied to the subject. It is therefore important to look at the pitfalls.

1. There is the high initial cost. This is composed of two components: the obvious purchase price and, not nearly so obvious but very important nevertheless, the hidden cost entailed in the initial inefficiency produced by staff having to use radically new methods and skills. The purchase price can be avoided by using a bureau service

but the inefficiency cannot be avoided in any way. It can be reduced by choosing a system with a good user interface which will reduce learning time.

2. There is the running cost. Like the initial cost this has both a visible and a hidden component. The visible component is the maintenance required on the computers, air conditioning and software. All of these can be reduced with careful management and good judgement to balance risk against cost, since all maintenance contracts are very much like insurance policies. The hidden component is the additional time required to manage the installation, which can also be minimised with care.

3. There is the effect on working relationships. A new technology is being introduced. It may lead the management to decide that it needs to bring in specialist knowledge in the form of a CAD/CAM manager. How does his role fit in with the role of the existing supervisor who no longer completely understands the tool he is using but has to retain the respect of his subordinates? What is his relationship with the supervisor? Does he advise or does he provide a service? What if the supervisor, being older, has difficulties in understanding the new tool and finds that some of his younger subordinates understand it better?

4. There is the danger that certain kinds of design cannot be done on the system because it lacks particular features overlooked during the initial evaluation. In this connection it is worth noting that there are many activities in the design process which cannot be done on computers at all, principally because the variety and nature of the data being manipulated is beyond current software. Clarke elaborates on this in Reference (1).

5. It is unfortunately possible to procure a software package that is always breaking down because of poor quality programming and testing or insufficient exposure to use in the field.

6. The following two features of a CAD/CAM system make it less suitable for early conceptual design. It demands precise numbers and the size of the screen prevents the designer from seeing the whole design at full scale, thus preventing him from using his judgement about sizes.

EXERCISE

Some years ago a CAD vendor used the slogan "We sell Productivity". Is this a good description of CAD? If not, what would be a better description?

Chapter 3 The graphics screen

GRAPHICS DISPLAY LIMITATIONS

The graphics screen is by far the most critical part of the whole CAD/CAM system. Computer hardware technology and the software techniques it supports are now well up to the task they have to perform on price and performance and both technologies continue to develop. Compared with the drawing board and the paper it replaces, however, the graphics screen still has deficiencies. To start with, it is so very much smaller than the drawing board. Figure 5.1 on page 44 shows a graphics screen in use. The size might be increased by some kind of projection system but if you did this you would not get a picture which was any easier to use, because it would be just a coarser picture. The problem is not in fact the size. The electronics which generates the picture on a display treats the screen like a rectangular array of squares to be filled in with black, white or a colour. Each square is called a *pixel* and is the smallest possible point which can be displayed. The number of pixels in the array is limited by fundamental characteristics of the circuitry. Current devices can achieve screens with about 2000 pixels from one side to the other and a corresponding number from top to bottom. Analysing an A0 sheet of paper in these terms, the smallest dot possible is about 0.2 mm across. The width of the paper is about 1180 mm so that there are about 5900 "pixels" from one side to the other, which means an A0 sheet of paper can hold three times the detail of a high resolution cathode ray tube (CRT) display.

But besides its better resolution, paper can be viewed comfortably in a wide range of lighting conditions, and the contrast remains high under all lighting levels and there is no glare. CRT displays, on the other hand, require carefully adjusted lighting levels and the glass envelopes they use catch reflections of ceiling lights and other bright surfaces so as to cause uncomfortable bright spots and glare.

As a consequence, one should not skimp on the graphics screen when selecting equipment. There are many software features one can do without but one cannot do without a good quality graphics screen. Obtain as high a resolution as possible and make a critical assessment of the quality of the picture, which can differ between screens of the same resolution. Picture quality is discussed later in the chapter.

RASTER CRT DISPLAYS

The raster scan CRT display is by far the commonest. It is actually a higher resolution version of the domestic television set and works in the same way (see Figure 3.1). A fine beam of electrons, the fundamental particles of electricity, is directed from the rear of the glass tube on to a phosphorescent coating on the flat screen at the front of the tube. A small bright spot of light is produced where it strikes the coating. Currents in coils round the tube bend the beam so as to make the light spot occur at different positions. By rapidly varying the currents the spot is made to move from one side of the screen to the other and trace out horizontal lines. At the same time, a gradual movement from top to bottom places successive lines under each other. The result is that the screen is completely covered with closely spaced horizontal lines from top to bottom. This way of covering the entire screen with the single spot of light is known as a *raster scan*.

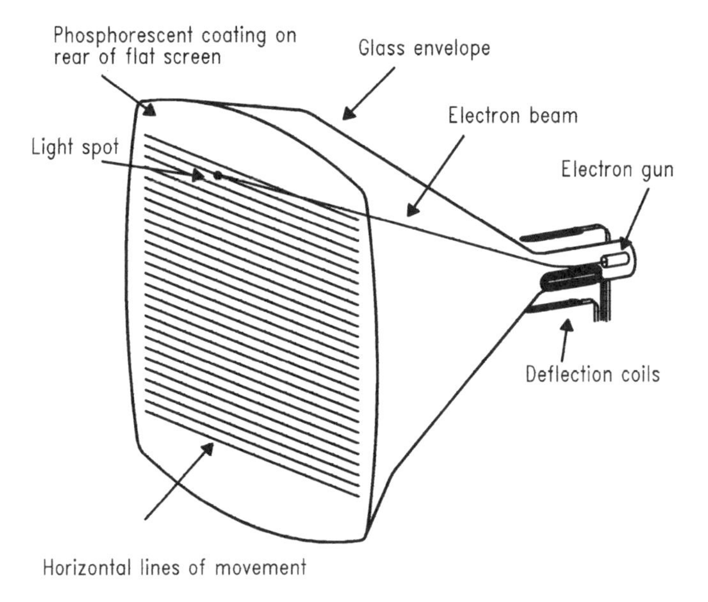

Figure 3.1 A cathode ray tube

The spot passes over each of the pixels of which the screen is considered to be composed and, as it passes, the intensity is instantaneously set to the required value for that pixel. The scan is repeated 50 times a second so that instead of seeing a moving spot of light the human eye just sees an illuminated screen with the pixels forming a picture.

Raster screens have one effect on line drawings which some users find disturbing when they first see it. No line is continuous, of course, because it is made up of a row of filled in pixels. For most lines this does not show, but when a line is close to the horizontal it seems to break up into a sequence of short horizontal lines looking like a flight of steps. The reason is that where the line crosses a horizontal scan line a short horizontal row of pixels is turned on. Then when it crosses the scan line below, another short horizontal row of pixels is turned on as shown in Figure 3.2. Designers used to producing sharp clean lines can be upset by the appearance but it is only the way the screen shows the lines. The actual data is quite accurate, more accurate than the best pencil line in fact.

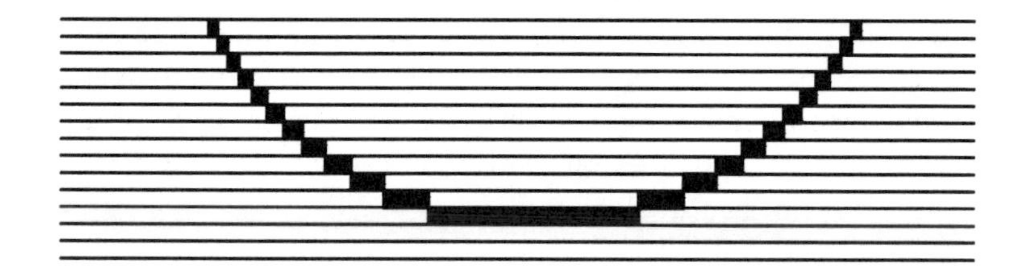

Figure 3.2 The effect of raster scan on a line

COLOUR CRTS

Colour displays use the same technique as domestic television sets to produce the colour. The phosphorescent coating is made up of three different phosphors, each of which produces a different colour. The phosphors are deposited in a regular array, fine dots of each colour being interspersed with each other. For each pixel there is a dot of each colour in a cluster, each dot activated with a different intensity. The dots are so fine and so close together that the eye mixes the three together to form a single colour.

The range of colours available is less than that of a television set: not because the tube cannot produce them but because the data storage required for the colour picture is limited. Television sets do not store their pictures. Instead, the signal received from the transmitter controls the intensity of the spot directly. Computers, on the other hand, put the data for an entire screen into a large memory known as a *display buffer* from which the display electronics gets its signal, and the larger the number of colours made available the larger is the memory required. As a compromise, some displays make a limited range of colours available at a time but then allow that range (or palette) to be selected from a much wider range by some kind of setting up procedure. This would not allow the palette to be changed in the middle of a picture but would allow it to be changed from one picture to another.

In order to reduce the display buffer size, computer displays also limit the number of steps through which the intensity of the light spot can be varied. The full range of intensity is covered but only a limited number of intermediate levels is possible.

Fortunately for engineering drawing, neither the limitation on colour nor on intensity range is a drawback. The use of even one additional colour is a major advance over the conventional paper technique. In fact it will take some time for designers to make good use of colour. As things stand at the moment, colour screens compensate for the limitation on resolution by helping designers to distinguish lines from each other and to assist in the control dialogue by indicating which lines have been selected for a particular operation. As time goes on, design offices will develop conventions for plotting out finished work in colour.

LIQUID CRYSTAL DISPLAYS

Liquid crystal displays consist of a thin layer of special liquid which changes its transparency according to an electric field applied across it. The layer is sandwiched between two glass plates which have transparent electrodes deposited on them to create the field. There has to be a separate pair of electrodes for every pixel. The very fine patterns required for the electrodes and the connections to them has made high resolution screens difficult but they could eventually replace CRTs An alternative to electrodes is the use of a laser beam focussed on the liquid to alter its transparency. The beam is deflected on to successive positions of the layer by mirrors. The liquid is made to keep the transparency set into it while the beam is off it. A very promising display using this method has recently become available. The screen is large, flat, less reflective than a CRT screen and has a higher resolution. Four separate small liquid crystal layers are arranged side by side with different colour filters. An optical system projects the four coloured images in registration on to the back of a large screen.

STORAGE TUBES

At one time the only practical display for CAD was the storage tube. They are no longer in production but there may be one or two still in use. Their resolution of 4096 pixels in both directions, which is much higher than the raster CRT, makes up for their drawbacks of a dim image, no colour and the requirement to redraw the entire image on the screen from time to time.

Instead of moving in horizontal lines the electron beam draws the lines on the screen like a pencil. A coating on the back of the screen temporarily changes state where the beam strikes. Electrons sprayed over the back pass through the coating where it has changed state and excite the phosphor on the front to display the path traced by the electron beam.

VECTOR REFRESH DISPLAYS

Vector refresh displays were the first to be used. They work like storage tubes without the storage layer which means that the lines have to be continually redrawn or refreshed sufficiently frequently for the human eye to perceive a steady picture. Their main problem was that there is an upper limit to the number of lines that can be drawn, beyond which drawing them takes so long that the display starts to flicker. Also, the signals used to deflect the beam are very irregular which creates problems for the circuitry generating them compared with raster scan tubes where the deflection signals are regular. Vector refresh displays are not used now.

INTELLIGENT DISPLAYS

The designer rarely displays the whole drawing on the screen. Usually, he has a magnified portion of it in front of him. The drawing therefore has to be scaled up and positioned before it can be displayed. As this is an operation which is local, self-contained and temporary it can quite conveniently be done by a small computer in the display itself. If the design is a three-dimensional model then it also has to be projected in some particular direction before it can be viewed, which is again an operation most conveniently done in the display itself. All this leads to the idea of an intelligent display incorporating its own computer which handles the whole business of converting the coordinates of the points in the drawing into the temporary coordinate system which meets the current magnification and position (or even projection) the designer requires. If the computer is fast enough the designer can be provided with knobs to turn to indicate position and magnification. Such a facility is of enormous value in overcoming the limitations in size and resolution of the graphics screen.

STEREOSCOPIC DISPLAYS

Stereoscopic films have been tried a number of times in the cinema over many years but without success. The reason is probably that our appreciation of solid shapes only partly depends on the stereoscopic effect. We also use movement either of the object or of our heads. The fact that we can manage without it, coupled with the requirement to wear special spectacles which all implementations to date have imposed, seems to be enough to prevent it ever becoming popular. This may change with the development of holography which does not require spectacles.

The increasing use of solid modelling in CAD/CAM has led to a feeling that it would be nice to have a truly three-dimensional picture on the screen, so one or two systems have been offered which provide this - using the inevitable spectacles. The systems work by obtaining a separate view of the model corresponding to the position of each eye. The views are then alternated in successive scans of the screen. The user wears special spectacles which switch the left and right lenses on and off in synchronism with the alternating scans so that each eye only sees its particular view. Two methods can be used to switch the lenses. In one, each lens is a liquid crystal element which can be made opaque or transparent by applying an electrical signal conveyed by a wire attached to the glasses. In the other, each lens is a piece of polaroid which will only transmit light polarised in a particular direction. A large liquid crystal element covering the face of the display switches the direction of polarisation of the light passing through it in synchronisation with the alternating scans. The scan for the left eye will therefore only be transmitted by the left-hand lens and that for the right eye by the right-hand lens. The glasses in this case require no electrical connection.

THE QUALITY OF COLOUR RASTER CRT DISPLAYS

The number of pixels in the horizontal and vertical directions respectively is often used to compare one display with another. However, these figures do not give an adequate indication of the quality of the picture. This depends on a number of other factors so that two displays with the same number of pixels can differ noticeably in picture quality. To understand these factors it is necessary to understand how a colour tube works. It will be recalled that a different intensity has to be applied to each phosphor dot

in a cluster according to its colour. The method used for selecting the dots of a given colour is known as the *shadow mask principle.* Three separate electron guns, one for each colour, are arranged close together in a row as shown in Figure 3.3. The beams they emit are made to converge on to the same spot on a perforated metal sheet located just behind the screen so that they pass through the same hole in the sheet together.
Since they converge on to the hole from different directions they will diverge as they leave the hole and strike the screen at different positions. These positions will correspond with the relative positions of the three guns. A phosphor dot of the colour corresponding to the beam striking it is located in each of the positions. Thus each gun will only activate dots of its own colour.

The dots are placed at a pitch of 0.2-0.3 mm and the perforated sheet (or mask) is about 0.01 inch thick and located about ¼ inch behind the screen. The holes in the mask must be held in correct registration with the phosphor dots. The mask is therefore fixed in a steel frame which is held in place on three bimetallic supports designed to compensate for dimensional changes occurring as the tube heats up. Registration is obtained during manufacture by fixing the mask in place before depositing the phosphors and then using it as an optical mask in a photolithographic process which etches the dot pattern in the three phosphors. The hole pattern in the mask itself is also made using photo-etching. It is important that the beam passing through the hole on to its intended dot does not hit any of the adjacent ones. The hole therefore needs to be slightly smaller than the dot.

During operation the mask can become magnetised and deflect the beam. To keep it demagnetised, a strong alternating magnetic field is generated momentarily by coils round the tube when it is powered up. It is for this reason that diskettes or other magnetic media should be kept away from colour displays otherwise data will be destroyed by the demagnetising field.

The three beams of electrons are deflected on to their destination on the mask by magnetic fields generated by coils round the outside of the tube. Two separate sets of coils are provided: designed to produce independent horizontal and vertical deflections. The beams are deflected through large angles to reach the edges of the screen. The horizontal coils must move the beams at constant speed in straight lines of equal length across the screen and the vertical coils must produce evenly spaced positions of the successive horizontal sweeps. In addition, the three beams must converge on to the same position on the mask whatever angle they have been bent through. Considerable attention has to be given to the shape and position of the coils and the waveform of the currents in them to meet these stringent conditions, and an enormous amount of research and development has gone into their design.

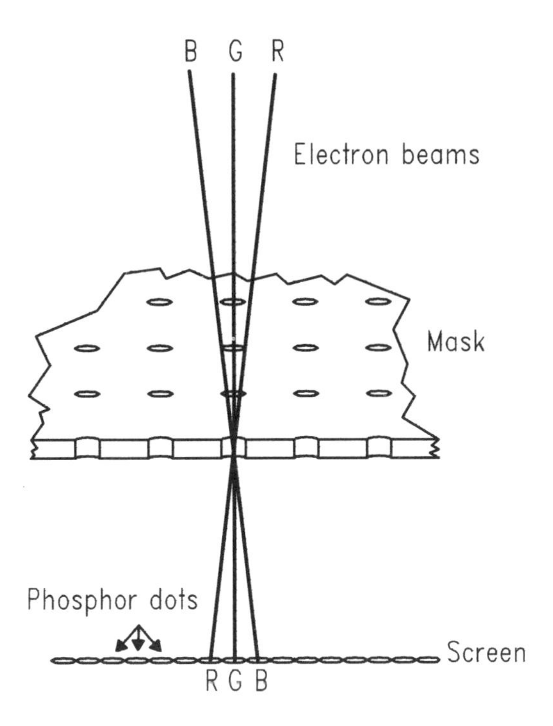

Figure 3.3 A colour display

As remarked earlier, when a display manufacturer quotes the number of pixels he is merely giving the number which the drive electronics can step through, or the number of storage locations in the display buffer. The picture quality depends on other factors as well. There is no point in having a spot so large, for instance, that adjacent pixels overlap. The size of the spot depends firstly on how finely the electron beams can be focussed down: the finer the focussing the finer they need to converge. Assuming the design of the coils and drive circuitry is good the production spreads in the coil geometry and circuit components have to be adjusted out in the final test of the display. This is an expensive activity as it involves individual attention to each display. Automatic adjustment methods are being developed and currently as many as six of the adjustments are made in this way using actuators to turn variable resistors in response to automatic optical sensing of test patterns on the screen. The variable resistors will eventually be replaced by digital systems in which the settings are written into small permanent memories.

Having got as small a spot as possible there is no point in going smaller than the holes in the mask, so these and the corresponding dots of phosphor behind them are a limiting factor. A number of problems combine to make reductions in the size difficult. The phosphor dot needs to be larger than the hole by a fixed margin. As the dot is made smaller the hole could become impossibly small. There are limitations in the chemical milling process used for making the holes and tighter controls will be needed in the manufacturing process to avoid holes being blocked by the phosphor powder. The photolithography used to make the dots will also become more difficult.

To sum up, the number of pixels, or addressability, is largely a matter of providing enough memory for the display buffer. The real quality of the picture depends on the width and linearity of the scan lines on the screen, which in turn depends on good design in the tube, deflecting coils and circuitry, and good quality control in the manufacturing process.

THE SAFETY OF CRT DISPLAYS

Anxiety has been expressed from time to time in the past over possible hazards in the use of CRTs. The first one was the danger of implosion. There is a vacuum inside the glass and the external air pressure could cause the tube to fail violently and project broken glass at the user. The danger is countered by making the screen about half an inch thick and providing a force which opposes that of the air pressure on the screen. This is done by making the screen slightly convex and then applying a compressive force on the circumference towards the centre which would act to make the screen even more convex, thus resisting the action of the air pressure to push the screen in. Regular tests are specified by regulations which require that the tube be struck on its own and in the cabinet and that the cabinet itself be struck. In all cases the glass must not travel more than a prescribed distance.

The other hazard which has caused anxiety is the possibility of radiation being emitted. The device used to generate X-rays is similar to that of a CRT since in both there is an electron beam striking a metal target, which is the way in which X-rays occur. The hazard is overcome by the use of lead glass for the tube and the amount of radiation which gets outside the tube turns out to be less than that from a piece of natural granite of the same size.

EVALUATING GRAPHICS SCREENS

In conclusion we can summarise the factors to take into account when evaluating CAD screens.

1. Obtain the maximum resolution (i.e. addressability). Physical size is of secondary importance to resolution. Resolution is a measure of the information the screen can transmit.

2. Assess the picture quality. For screens of the same resolution the quality of the picture can differ depending on such factors as spot size and colour convergence. Examine picture quality over the whole screen.

3. Look for facilities which make the most efficient use of the display area such as:

   - Instantaneous pan and zoom
   - Instantaneous 3D rotation
   - Multiple windows on to the CAD drawing
   - Minimal area occupied by screen menus or menus which do not use the screen at all

4. The flatter the screen the less are the problems of reflections from lights behind the user.

5. The height and tilt of the screen should be easily adjustable.

EXERCISE

How can a screen be of a poor quality yet have a high resolution?

Chapter 4 Interactive control devices

Having discussed the devices which make the drawing visible to the designer we now move on to the devices the designer uses to create and alter the drawing. The drawing is actually a bank of data in the computer memory and the devices are the means by which the designer tells the computer what to do to the data. Nevertheless, they must be so natural to use that the designer feels he is using them like his conventional drawing instruments.

POINTING DEVICES

The user of a CAD system must be able to point to items on the graphics screen and to indicate where he wants various items placed. The earliest device was the light pen. It was used with the vector refresh screens and consisted of a pen-like probe with a light detector in it which the user held against the screen at the desired position. The light detector sensed when the spot on the screen passed the end of the probe and sent a pulse to the display electronics which then registered where the spot was at that instant. Although at first sight it seems the logical and natural way to indicate position, the light pen has a number of disadvantages which is probably why it is no longer in regular use. Firstly, it cannot work at all on blank parts of the screen because there is no spot to pick up. Secondly, it is difficult to aim because, being rather thick and not pointed like a pencil, it obscures the part of the picture one is trying to aim it at. Thirdly, the user has to hold it up against the screen which is inconvenient. Having said all this, it was nevertheless used successfully for many years, various methods being used to overcome its problems.

Current displays all use a screen cursor. This consists of a cross displayed on the screen, as shown in Figure 4.1, which is moved around the screen using various control devices. The value of having a screen cursor is that it is part of the picture on the screen so that there is no difficulty in seeing its position in the drawing. Although it is part of the picture the display electronics generates it separately so that it can be moved about on its own.

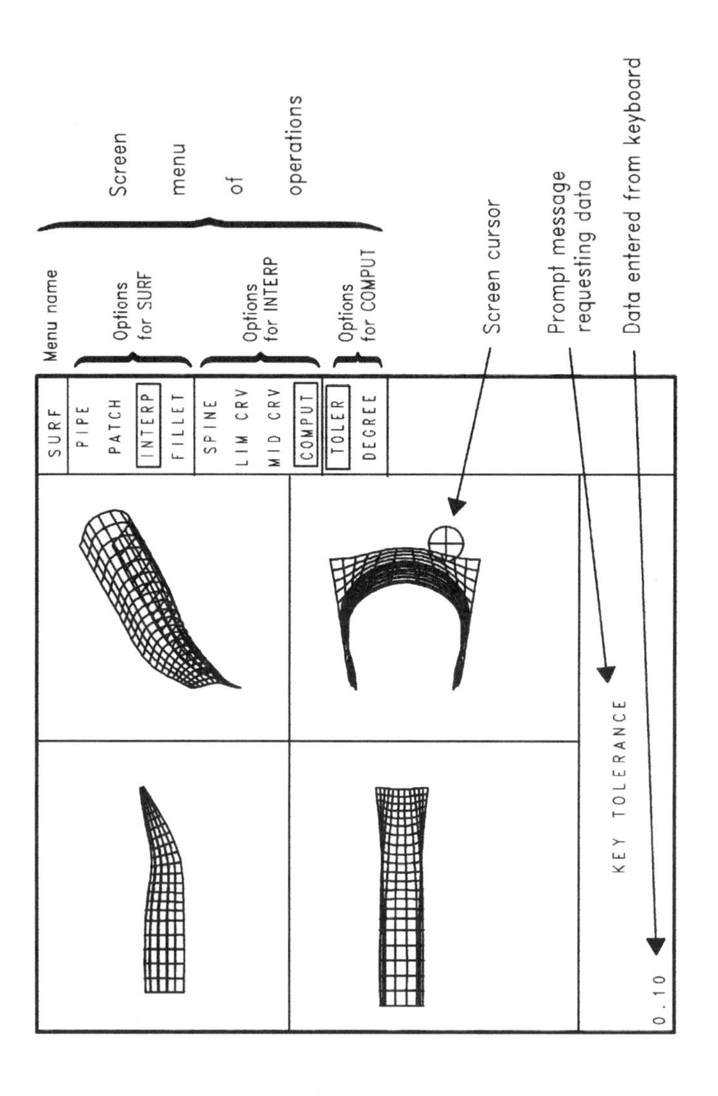

Figure 4.1 A CAD screen of a surface design with a screen cursor

Various cursor control devices are employed to move the cursor about. The commonest control device is the graphics tablet or digitiser. This is like a small drawing board. The user holds a pen-like probe over it as if writing (Figure 4.2). Its coordinates are picked up by the electronics and relayed to the display so as to make the screen cursor follow it. As an alternative to the pen, a puck (Figure 4.3) is sometimes used which is a small flat circular frame holding a glass window with cross-wires marked on it. A related application of the graphics tablet is in copying drawings into the CAD system from paper. The drawing is fixed to the tablet and the positions of the vertices and key points on the drawing are entered by locating the cross-wires over them. This process is known as *digitising*. Before digitising a drawing, a calibration procedure is necessary to set the position of the origin and the scale of the drawing.

A device which has become popular recently is the mouse (Figure 4.4), which has some similarity to the graphics tablet. A mouse is an object looking rather like its namesake which the user slides around on the surface of the table. A ball or two wheels in contact with the table underneath revolve as it moves. Sensors detecting the revolutions communicate the movement to the computer via a cable. The mouse usually has one or two buttons on it which the user can press to initiate an action. Since the movement of the mouse cannot be sensed while it is moved out of contact with the table it does not indicate absolute position with respect to a fixed origin as a puck on a graphics tablet does. A mouse is therefore only used for moving a screen cursor about and not for digitising a drawing. It is not fitted with any pointer or cross-wires. The ability to lift and move a mouse without it registering anything is an advantage where the table area is limited.

A very different cursor control device is the joystick (Figure 4.5) which "flies" the cursor around the screen. It is literally a stick projecting up from a box which can be tilted in any direction. The cursor moves in the direction of the tilt at a speed proportional to the deflection, thus allowing the user to fly the cursor rapidly for large movements or slowly for small accurate positioning. A spring biases the joystick back to the central stationary position when it is released.

Last but not least there is a device which has been used successfully for years on air traffic control and military radar screens but which does not seem to have caught on in CAD. This is the tracker ball. It is a large ball positioned in the control panel so that a small part of its circumference projects through (Figure 4.6). It is free to rotate in any direction and the cursor follows its movement. The ball can be positioned very finely for accurate location and with a reasonably large mass and low friction it can be spun to give fast movement over large distances.

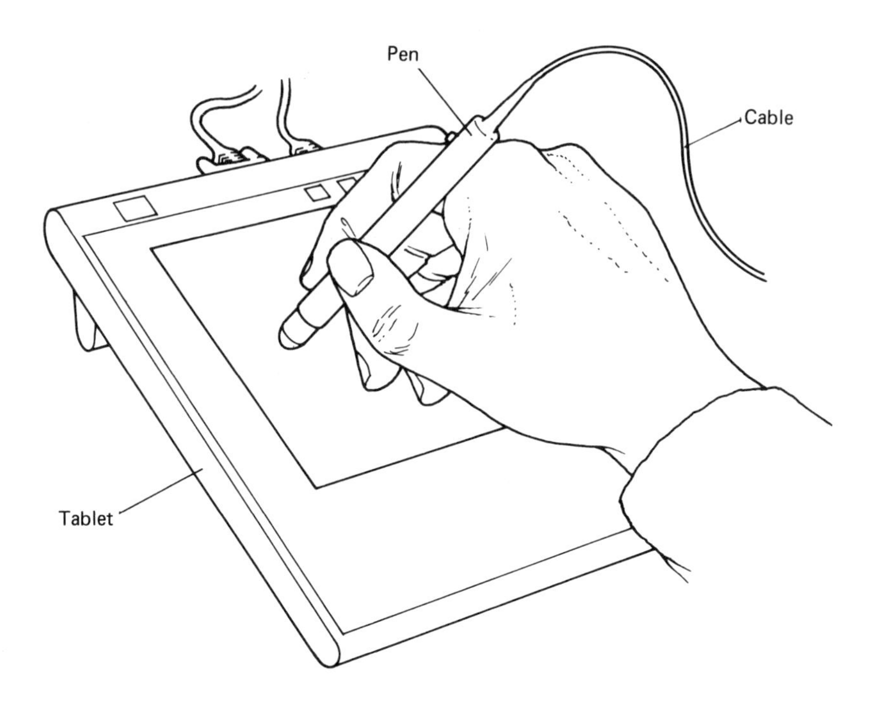

Figure 4.2 A graphics tablet and pen

Before we leave the subject it is worth noting that Personal Computer CAD software often provides for cursor control using the cursor control keys on the keyboard. This method is far too slow for serious work.

THE CONTROL DIALOGUE

The user needs some way of telling the computer what he wants done to the drawing. It should be as quick as possible to learn and be capable of communicating a large number of possible operations. These two requirements are in direct conflict since the larger the number of possible operations there are the more there is to learn. The control method should also be capable of fast communication for the experienced user.

The control of a CAD/CAM system is extremely complex. Not only is the number of possible operations enormous but numerical data and screen positions have to be supplied as well. A communication problem of this complexity requires many of the features of ordinary human communication such as question and answer, feedback from the listener to signal if

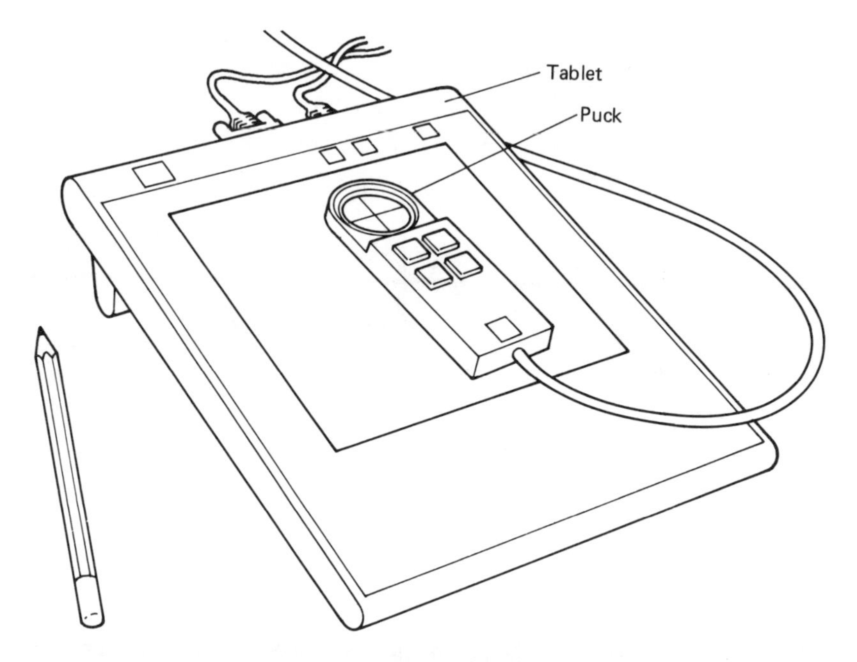

Figure 4.3 A graphics tablet and puck

he has not understood etc. Communication between user and software becomes a dialogue.

The dialogue has many of the characteristics of language even though nothing is spoken or written. In fact, the control of even simple machines can be analysed in terms of language. The vocabulary of a hot drinks machine, for example, is the set of buttons while the requirement in many machines that the "sugar" button must be pressed after the tea or coffee selection button corresponds to a simple rule of grammar. Looking at them you may have noticed that in some there is a separate button for each combination of tea/coffee, black/white and sugar/no sugar so that you just press one button out of eight to get the drink while in other machines there might be three pairs of buttons and you press one in each pair to make your selection. The former has a vocabulary of eight words and no grammar while the latter has a vocabulary of six words and a rule of grammar saying that you must press three buttons in a particular sequence. The latter is a more complicated language with nouns (tea and coffee) qualified by the adjectives black, white, sweetened and unsweetened. The use of adjectives has reduced the vocabulary by two buttons in eight. If extra sugar was an option then it would be reduced by five buttons in twelve (2 + 2 + 3 instead of 2 x 2 x 3). The use of qualifiers or parameters is essential in any complex language.

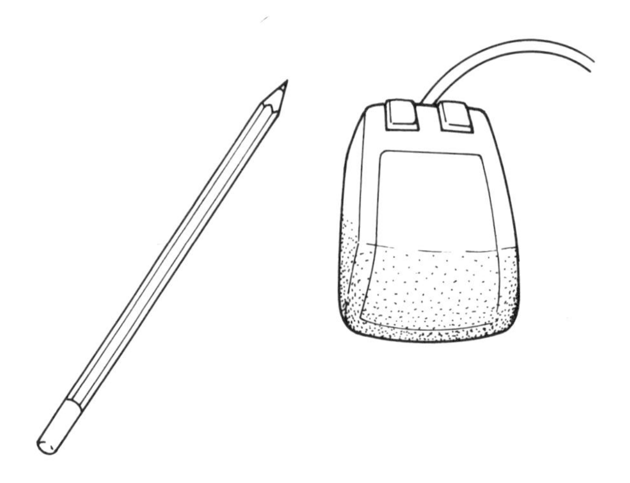

Figure 4.4 A mouse

These considerations can be seen in the design of the control methods of a CAD/CAM system. Nearly all commands are parameterised to a greater or lesser degree according to the taste of the designers, depending on whether they prefer a large vocabulary and shorter sequences of commands or a small vocabulary and longer sequences. In any case, numerical and positional data will be parameters. For example, the action might be to draw a line and the parameters of the action might be the coordinates of the end point, the length and the direction.

An important feature of a good dialogue is to allow the user to terminate an interaction before it is complete. Most good dialogues also allow certain commands (e.g. zooming the display) to be used in the middle of another command before the interaction is complete.

Having looked at some general considerations in setting up communications between the designer and the computer let us look at how it has been done in practice to date. The dialogue requires both input and output devices.

THE ALPHANUMERIC CRT

The alphanumeric CRT or VDU (visual display unit) is an output device most people will be familiar with since it is the usual device attached to a computer. As mentioned in "The graphics screen" on page 21, the graphics screen is a critical device. Since the control dialogue is only carried out with words, the use of a separate cheaper device restricted to characters to display messages for the user avoids using up valuable graphics screen

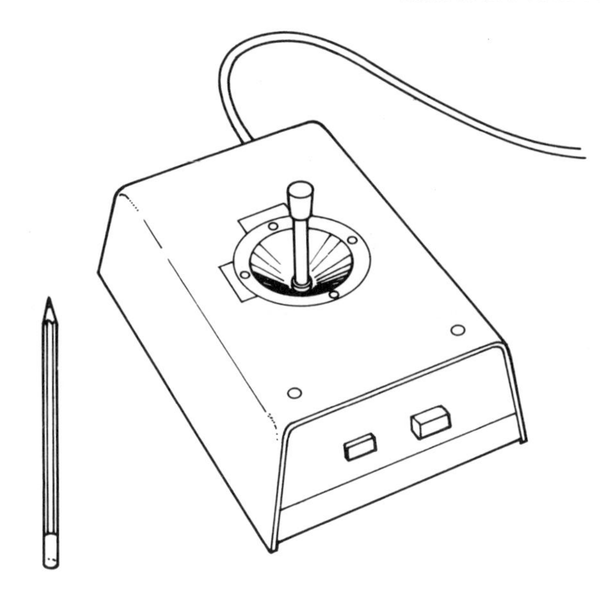

Figure 4.5 A joystick

space. Also, many of the messages are subsidiary to the task in hand and are best kept away from the main area of attention.

THE PLASMA PANEL AND THE LIQUID CRYSTAL DISPLAY

These output devices which have been developed in recent years are alternatives to the alphanumeric CRT. They fit into a smaller space than the CRT and can be made in versions displaying only a few lines which are cheaper and very much smaller than CRTs. However, both types of display are monochromatic which may be the reason why they have not been very extensively used.

SCREEN MENUS

The screen menu is the easiest control technique to learn. The software puts a list of operations on the screen as shown in Figure 4.1 and the user picks one of them by positioning the cursor. Alternatively, the list is numbered and he hits the corresponding number on a keyboard. The method will be familiar to users of Personal Computers. If the list is put on an auxiliary alphanumeric screen, more screen area is made available for the vital graphics display but the auxiliary screen will need to be placed close to the graphics screen to reduce fatigue arising from the user

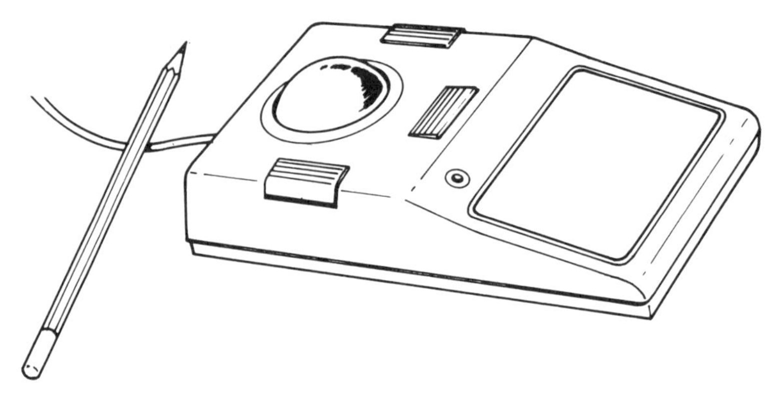

Figure 4.6 A tracker ball

constantly shifting his gaze from one screen to the other. No more than about ten operations are shown at a time, not just because of space on the screen but because a long list increases the information the user has to read and slows him down. To handle the large number of operations available, a large number of menus are provided arranged in a family tree or hierarchy. There is a top-level menu, each item of which brings up a second menu. Sometimes this is carried on to a third or fourth level before an actual operation is selected. Quite often the top level is not a menu but a small panel of buttons known as a *function keypad, programmable function keys* or *pfks*, shown in Figure 4.7.

The advantage of screen menus is that you always know what choices are available to you so that they are easy to learn. A weakness is that the process of finding and selecting the desired operation is slow. To avoid this it is important for the software to allow the experienced user to make his selections without having to wait for the menus to appear. An unexpected disadvantage is that you can get lost in the hierarchy. There should be a way of getting back to the top if you find that you have ended up with the wrong menu as a result of making a wrong selection. To avoid confusion the display should always show the sequence of selections which led to the present position in the tree. It is also best that the tree has no more than three levels.

TABLET MENUS

You may remember that a graphics tablet is a board which can sense the position of the puck or pen placed on it. In a tablet menu the surface of the tablet is marked out with boxes, each of which represents an operation. The pen has a switch in it to detect when it is pressed down or the puck

Figure 4.7 A function keypad

has a button. To pick an operation the user presses the pen on the desired box or positions the puck and presses the button. The coordinates of the pen or puck are sent to the computer where the software looks up the corresponding operation in a table. A large number of boxes can be marked on the tablet and by using removable overlay sheets or a book many more can be provided. The software has to be informed which sheet or page is in use, of course. Often a tablet menu is combined with cursor positioning by allocating a portion of the tablet for cursor positioning and the remainder to menu boxes. Such a tablet menu is shown in Figure 4.8.

The advantages of tablet menus are that an operation can be initiated with one action and the overlay sheet can be printed with coloured blocks and other additional material such as notes or a flow chart to help the user find his way around. Because they do not lead you through the sequence, training is essential. Some systems provide a means for the user to redesign the menu.

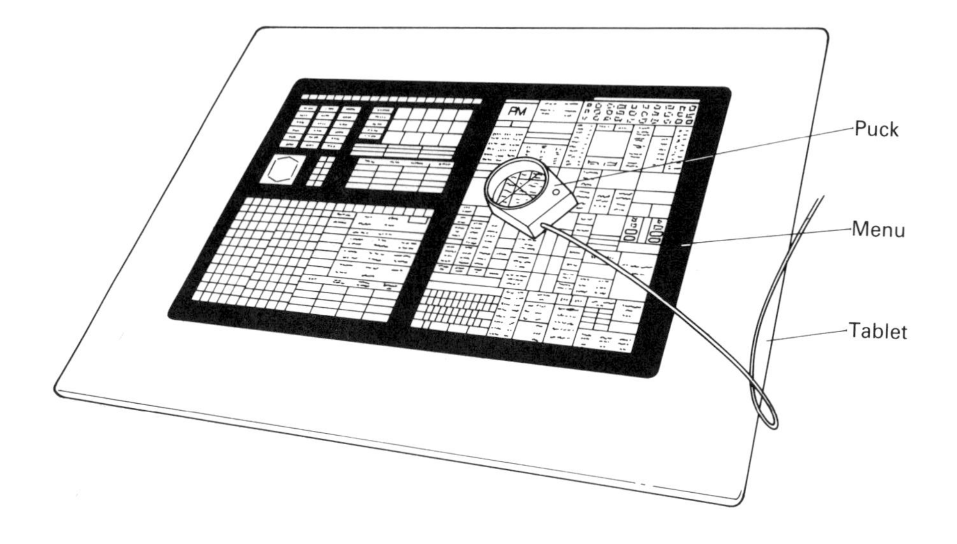

Figure 4.8 A tablet menu and puck

STROKE PATTERNS

One supplier has been successful with a system in which the user makes a stroke of a particular shape with the pen in the cursor control area of the graphics tablet. It might be a rough circle enclosing part of the drawing, for instance. The software interprets the shape as a command to be applied to the part of the drawing enclosed. There is a repertoire of strokes covering many of the commonly used operations.

VOICE INPUT

This has been offered. Voice input devices have to be trained for particular individuals. Steps would have to be taken to avoid the noise from one user distracting his neighbours.

DATA INPUT

A conventional typewriter or computer keyboard is usually provided for the user to enter text and numbers. Sometimes a typewriter keyboard is drawn on the tablet menu. There have been systems, even, which provide a set of boxes in alphabetical order on the menu tablet, but such is the entrenched position of the dreadful QWERTY keyboard, they are rare!

THE WIMP ENVIRONMENT

A realisation in the late 1970s that the reducing price of electronics would result in computers becoming widely available clerical "tools" caused the Xerox Corporation to form a Computer Science Laboratory in Palo Alto, California just a few miles from Stanford University. The work of the laboratory produced, among other things, the Ethernet Local Area Network and a small Personal Computer known as the Alto. New user interface software was developed which eventually had a major influence on the subsequent Macintosh computer manufactured by Apple, the Sun workstation and the Apollo workstations.

The user interaction method has come to be known as "WIMP", standing for Windows, Icons, Mice and Pop-up menus. The system models a desk top on which are piled sheets of paper. Usually, the sheets are not stacked exactly on top of each other so that a part of each sheet is visible at the edge of the one on top. The sheets are manipulated by a mouse. Picking the edge of a sheet brings it to the top and its position on the screen is altered by moving the mouse. Each sheet is in fact a window on to a file or a program so that several files can be viewed at once or several programs run at once. To reduce the space occupied on the screen, a window can be converted into a small graphical symbol known as an *icon*. Picking the icon with the mouse expands it into a window again. Since a screen menu is only needed when a command is being given it is made into a window which appears only when required. A common technique is to provide icons for each menu which are opened up by selecting with the mouse.

This user interface method is likely to become universal. A standard system for implementing windows on networked computers, X Windows, is described in "CAD data exchange standards" on page 175.

EXERCISE

Choose the set of control devices you would like to use, giving your reasons as fully as possible.

Chapter 5 Workstation layout

In the previous two chapters we have introduced the graphics screen, a possible subsidiary alphanumeric screen and the various control devices. A particular set of these will be available for use by the designers depending on the particular make of CAD/CAM system installed. The designer will sit at a table or desk on which these devices have been installed. This collection of display screens, interactive control devices and furniture, being more complex than a typical computer terminal or personal computer, is usually known as a *workstation*. A workstation may also include a computer running the CAD software, thus making it almost completely self-contained. Increasingly, the term "workstation" is being used to mean just such a self-contained arrangement. The matter is discussed further in "The computer configuration" on page 59. In fact the difference between the two meanings is not significant in most of the places where it is used in this book and no distinction will be made between them.

The term is very apt since, if you are at all serious about CAD/CAM, the workstation is going to be used intensively to produce a lot of work. An individual user will be sitting at it for long periods.

A typical workstation is shown in Figure 5.1. The physical arrangement of the various items will greatly affect the degree of fatigue the user experiences. While sitting in a comfortable position he must be able to reach the various control devices without stretching and to operate them comfortably without strain. At the same time the user must be able to see the displays and the control devices clearly. Quite a number of conditions have to be fulfilled at one and the same time to achieve these objectives. Firstly, the physical positions of the control devices and screens must be correct in relation to the eyes, hands and arms of the user, which means correct positioning of control devices, screen and chair seat. Because of the range of sizes of likely users, any particular arrangement which is good for one person may be bad for the next, bearing in mind that the present cost of workstations means that they will usually have to be shared. This means that it is essential to make the position of everything as adjustable as possible.

By connecting them to the workstation control unit with flexible cables the various control devices can be easily repositioned to suit individual operators. Furthermore, they are usually designed to be comfortable to operate when simply placed on the table so that the only adjustment needed is their horizontal location on its surface. The display screens and the chair

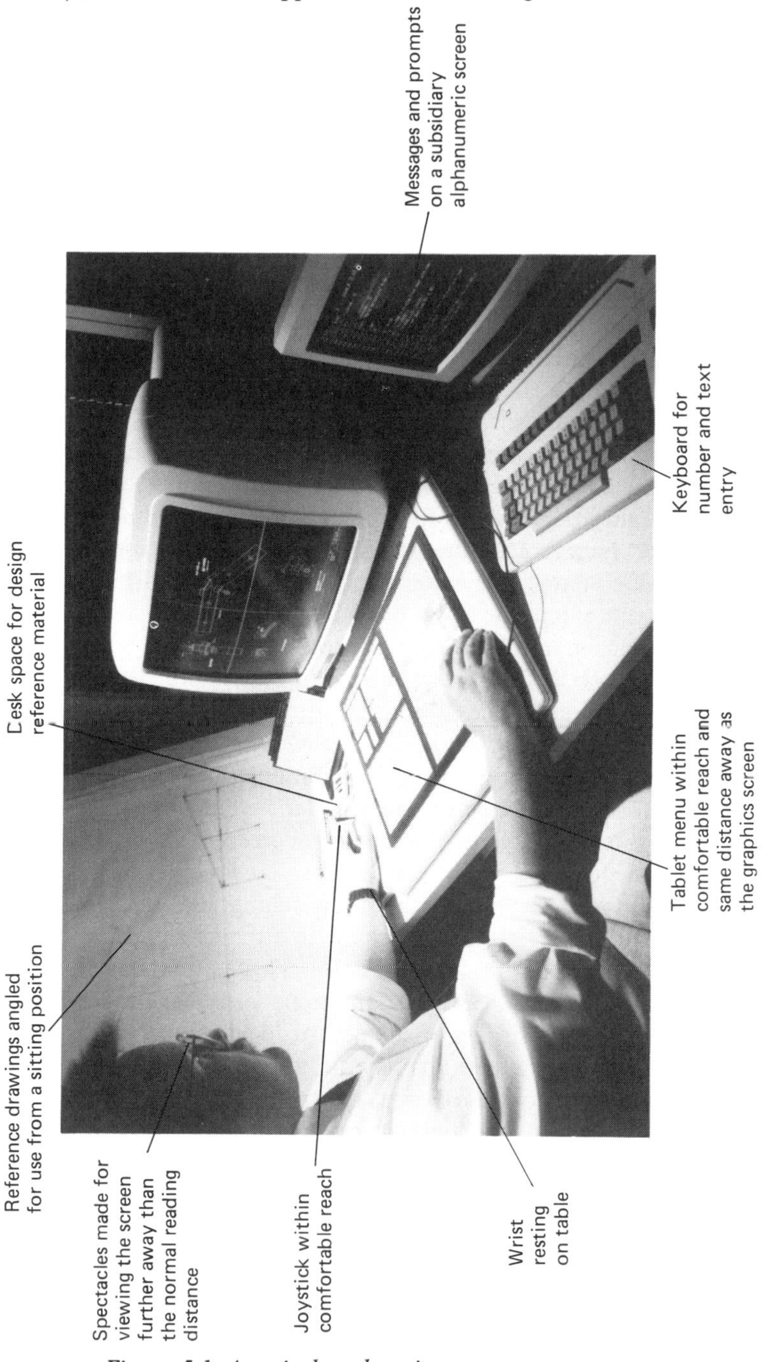

Figure 5.1 A typical workstation

seat, on the other hand, are heavy and have the additional need of adjustment in the vertical direction. Adjustable chairs are now commonplace in offices so should present no problem with the possible exception that the range of adjustment should be adequate.

The positioning of the display screens, however, usually requires quite some attention. There are two visual factors which have to be taken into account. The less obvious of the two factors, but one which can present problems difficult to trace, is the need of the eye to focus on the screen. Present screens need to be placed a little further from the eye than printed paper, which means that someone with glasses designed for normal reading may not find the screen properly in focus and suffer from eye strain. It may be necessary to instruct his optician to provide for a near point at the screen distance which will be further away than usual. Another consequence of eye focussing can occur because the user is continually changing his gaze between screen and menu tablet or keyboard. If they are at different distances from him, his eyes are continually having to refocus.

The second of the two visual factors is often very apparent. The glass of the screen is a mirror which reflects back everything in front of it to the user. Not only that, it is a convex mirror which reflects objects from a wider area than a normal flat mirror. Each reflection is an optical image located in space further away from the user than the picture on the screen he is looking at, with the consequence that it is duplicated by parallax as well as being out of focus. The objects which cause the most trouble are ceiling lights and windows. These are bright enough to obscure whatever part of the screen they appear on. The user is not fully conscious of them because they are out of focus but they still obscure the screen and his efficiency is reduced. The best way to avoid the problem is to locate each user with his back to a wall. Other solutions are diffusers on ceiling lights specially designed to project light downwards, ceilings lit by uplighters, curtains hanging in front of the lights and false ceilings composed entirely of light-diffusing panels.

If you are using subdued lighting, make sure that the level of illumination of the area the user sees beyond and around the screen matches that of the screen picture and that of the control devices and other things he may be looking at. This avoids the eye continually having to adapt to differing levels as the user transfers his gaze from one part of his surroundings to another.

Finally, most workstations at the time of writing seem to be designed on the naïve assumption that designers are not going to use conventional paper drawings. Next to every conventional drawing board there is nearly always a reference table big enough to take an A0 drawing with space to spare. This should be the case for the workstation - with one difference.

The designer always sits down when using a CAD workstation so an angled reference table is needed instead of a horizontal one.

But besides all the paper the designer needs in order to do his work the CAD/CAM system requires additional amounts of paper in the form of manuals on how to operate the equipment. At least some kind of quick reference guide will need to be on hand, if not the whole user guide or commands reference manual. So a place will have to be found for these as well.

At the time of writing, workstations in general are very poorly designed. The aim of this chapter has been to present the factors which need consideration in laying out or adapting the workstation to make it efficient, comfortable and less fatiguing to use. The following steps are suggested for getting the layout right:

1. Solve the lighting problem by positioning operators with their backs to the wall or installing suitable ceiling fittings.

2. Choose tables and chairs which allow adequate adjustment of the positions of the various components for comfortable use.

3. Choose additional units for reference tables and stowage of manuals.

The design of workstations will be greatly influenced in the future, particularly in Europe, by a new European Council directive on the subject (see Reference (40)).

EXERCISE

If someone complained to you that he found using a particular CAD workstation tiring, what questions would you ask him?

Chapter 6 Printers and plotters

This chapter deals with the various devices available for putting drawings out on to paper, tracing paper or polyester film.

WHY PLOT OUT?

Before we deal with the devices available it is worth asking the two questions: why and when should you plot out? Obviously, the most important use of plotted output is in issuing the finished drawings for manufacturing. As the technology develops there will be a steadily increasing number of places where the Production Department and, in certain industries, subcontractors will have to accept the drawing in its electronic form. When numerically controlled machine tools are used the electronic form will be converted electronically into codes for directly controlling the machines. Graphics screens in the workshop will allow people to look at drawings in their electronic form. However, paper is such a convenient medium in so many ways that one cannot imagine its complete elimination. In considering the replacement of paper by electronic screens it is important to distinguish between paper as a working medium and paper as a medium for conveying information to humans. The CAD/CAM system is successful because it is superior to paper as a working medium.

Related to issuing drawings for manufacture is issuing drawings for subcontractors to tender against. Subcontractors are even less likely to accept drawings in electronic form than the Production Department. Where they have a CAD/CAM system it will have to be compatible with the one used to produce the drawings before it can accept them in electronic form.

A meeting is one occasion when good looking drawings are important, particularly if clients, customers or higher management are going to be present. Graphics screens are badly suited to large audiences unless they can be projected without loss of resolution or quality. Well plotted paper drawings will therefore have their place in technical presentations for a long time.

The place where plotting comes into its own is in preparing sales proposals. In fact the whole of CAD is particularly effective in this area. By taking previous designs and modifying them using the CAD/CAM software the designer can produce proposal drawings quickly. The plotter can then output them to a very high standard. The ability to plot in colour can also be usefully exploited.

It might be thought that an important use of the plotter is to produce the finished version of the drawings for storage. But is there a need to keep archival copies of drawings on polyester film any more? The archival copy is now the computer file on an archive tape or disk. That is what will be used for developing the new issue of the design.

At first, some beginners will feel a need to plot out frequently because they are accustomed to having drawings on paper at hand. Although that feeling will wear off there will always be a need to have drawings at the desk for reference and marking up. They need not be of high quality and often only a portion of a drawing is needed. For this purpose, screen copy or "hard copy" devices are provided as additions to workstations. They pick up the picture on the screen, directly from the screen buffer. The quality is no better than that of the screen but they are quick and, also, allow the user to select the portion of the drawing he wants to take away with him for reference.

The use you will be making of plotted drawings is an important factor in choosing both the plotter and the media it is to use. We will return to the subject when we have considered the various devices available.

PEN PLOTTERS

The pen plotter was the earliest device to be used by computers for outputting drawings. It works by mechanically moving a pen over the paper. The simplest mechanism works rather like a gantry crane. The pen is moved from side to side across the paper in a carriage running along a straight track which bridges the paper like a gantry. To move the pen up and down the paper the gantry is moved on tracks at each side. This arrangement is known as a flat bed plotter and is shown in Figure 6.1.

In models handling the larger sheet sizes the gantry remains stationary and the paper is moved by various means. A popular method is to drive the sheet back and forth by a friction wheel gripping it at each side as shown in Figure 6.2. This is a simple technique which makes it easy to load the paper at the start of a plot. A minor disadvantage is that it is not possible to draw right at the edge of the paper. Another arrangement, the drum plotter, fixes the sheet round a drum which rotates back and forth under the gantry. This is shown in Figure 6.3. In another variation the sheet of paper is fixed to a flexible belt. At the top the belt passes over a roller beneath the gantry and at the bottom over an idler roller.

The paper can also take the form of a continuous strip or web passing over a roller situated beneath the gantry. Holes punched along each side

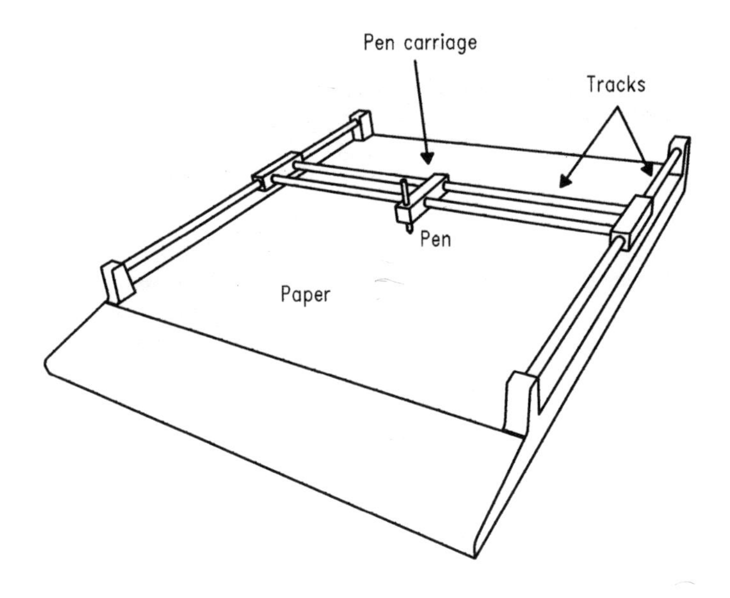

Figure 6.1 A pen plotter - flat bed type

of the paper engage with sprockets on the roller which propels the paper back and forth under the pen. The accurate positive control of sprockets is needed to avoid skewing which can accumulate over a long length. The paper is fed from a roll and taken up on a second roll. To maintain tension and provide a buffer between the drive roller and the take-up and feed rolls, a loop of paper is pulled down on each side of the drive roller by vacuum chambers as shown in Figure 6.4.

There are significant differences between using a plotter with individual sheets of paper and one with a continuous roll of paper. Each has advantages and disadvantages as summarised in Table 6.1. A plotter with a roll can work through a large batch of drawings relatively unattended on its own whereas with cut sheets each time a drawing is needed the user has to go to the plotter and put a new sheet on it. On the other hand, rolls have to be specially manufactured to suit the plotter whereas any piece of paper can be used on a sheet-fed plotter, thus allowing a wide choice of paper and the use of pre-printed paper. Although roll-fed plotters do not require the user to load the sheets on to them, work still has to be done to cut up a roll into individual drawings once it has been plotted. Also, changing the plotting medium requires stopping the plotter, removing the current roll and mounting a new one. Because both forms have advantages, plotters are now available which can be easily converted between roll and sheet feeding.

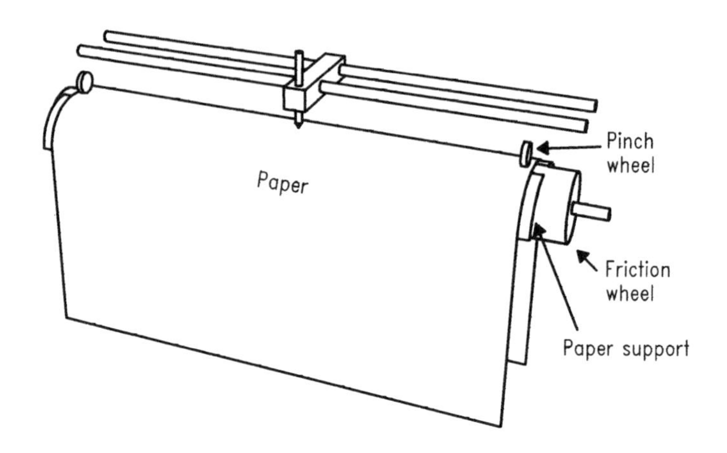

Figure 6.2 A pen plotter - friction wheel type

Table 6.1 Comparison of sheet- and roll-fed plotters

Sheet-fed plotters	Roll-fed plotters
Wide choice of material	Limited choice of material
Pre-printed material possible	Pre-printed material expensive
Rapid change of material	Can work unattended
Each drawing requires attention	Each roll requires attention

We will now turn to the vital subject of pens. These seem to consume more time and effort than anything else in the CAD/CAM system. All pens are simply better engineered versions of ordinary writing or drawing pens without losing the drawbacks. These are the characteristics of currently available pens:

Fibre tips — Although cheap and reliable, they lack the fine point desirable on engineering drawings.

Pressurised ball points — Cheap and reliable with a long life. Their only possible drawback is their somewhat low density of line compared with other pens.

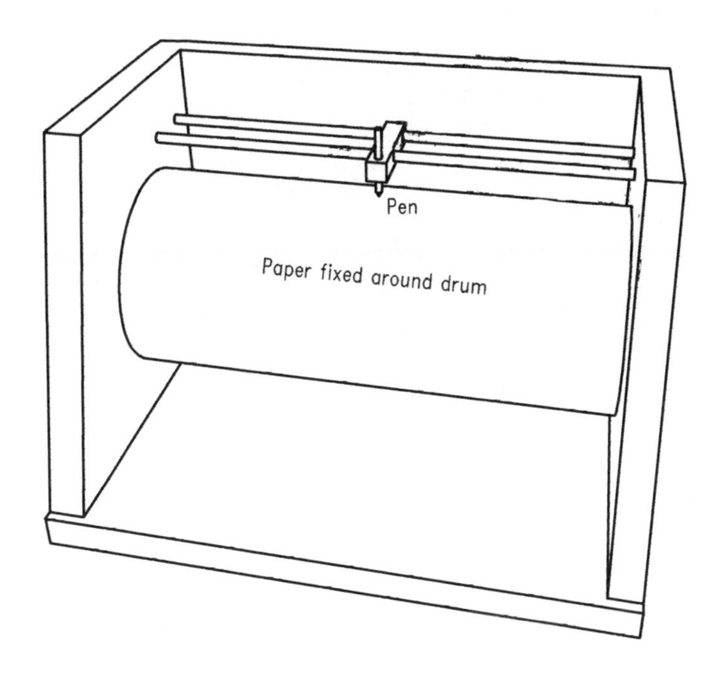

Figure 6.3 A pen plotter - drum type

Liquid ink ball points

More expensive with a shorter life than pressurised ball points. The line density is superb, matching that of ordinary ink pens. They have a tendency to miss occasionally over certain patches of the paper.

Draughting pens

Their main value is the high density of line on polyester film and the precise control over line width. These advantages are almost overruled by their habit of clogging up given the slightest chance.

Ceramic tips

These are similar to draughting pens but are supplied as a sealed disposable unit. They are more expensive than draughting pens but more reliable.

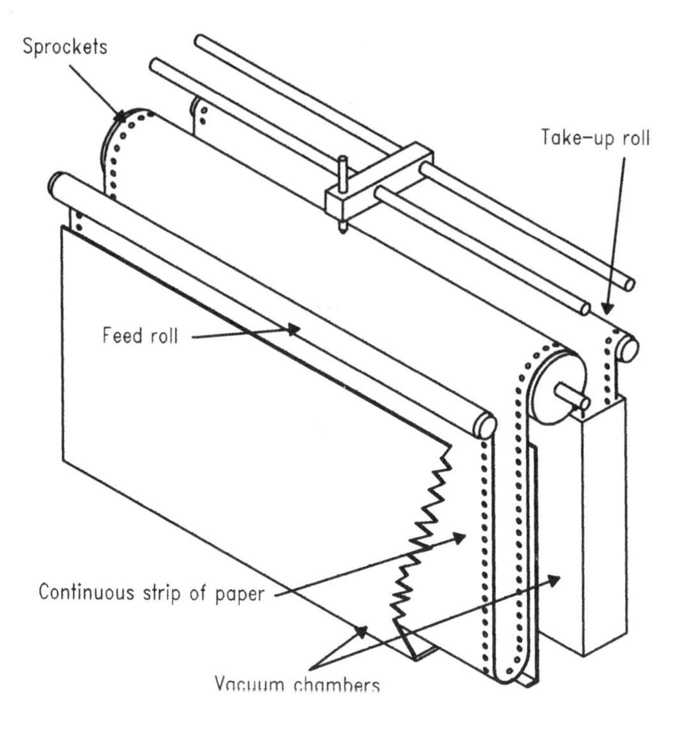

Figure 6.4 A pen plotter - roll type

The best rule is to use pressurised ball points for day-to-day work unless the high density line of the liquid ink ball point is preferred. It is possible to make dyeline prints off liquid ink ball point plots on lightweight (90 gm) cartridge paper.

If you have to produce polyester film plots for some reason then use draughting pens or ceramic tips. Draughting pens require a lot of care and attention. Procure and use regularly both an ultrasonic cleaner and the special magnifier for examining the tips. Tips wear quite rapidly and the wear cannot be seen with the naked eye. The ink is also critical. Besides flowing well, it must adhere to the film, not dry so fast as to clog in the pen yet not dry so slow as to transfer when the film rolls up. Watch the shelf life which is usually surprisingly short. You may have to make trials of various ink/pen/film combinations in order to get satisfactory results. The chance of pens clogging up is reduced if the plotter automatically puts a cap on the pen as soon as it has finished with it. It is also reduced by using only one pen so that it is in continuous use. This is possible if the plotting software is able to thicken the lines by making several strokes a controlled distance apart over each line.

The choice of medium for pen plotters deserves a little thought particularly if a roll-fed plotter is in use. Vellum is inclined to tear although it is easy to make dyeline copies from it. Polyester film is strong, good for dyelining but expensive and really needs draughting pens for best results. Where there is a heavy load of printing such as in issuing drawings for subcontractors to tender against, some prefer to plot on film or vellum and dyeline simply because the latter is faster even though it requires more labour than a roll-fed plotter. The author's preference for medium is lightweight 90 gm cartridge paper. It is strong enough for use in the workshop, has a smooth white surface which makes the drawings look good and is not too expensive. If you use liquid ink ball points you can even dyeline print from it with the exposure turned well up.

ELECTROSTATIC PLOTTERS

Electrostatic plotters use a similar principle to that of photocopiers. The basic arrangement is shown in Figure 6.5. The paper is given an electric charge wherever there is to be a mark. It then comes into contact with a fluid containing small black particles of toner at the toner head. The charge attracts the particles of toner and the paper emerges with black particles adhering where the paper has been charged. When the fluid evaporates the toner is permanently fixed to the paper. To put the charge on the paper in a pattern corresponding to the pattern of marks required, the paper passes under a long line of little wires or "nibs" stretching across it. Where a mark is required the control electronics puts a pulse of high voltage on the appropriate nib at the right time.

The way in which electrostatic plotters work, being very different from the way pen plotters work, is responsible for their particular advantages and disadvantages. Electrostatic plotters have the advantage of few moving parts and no pens to clog up or run out, so are inherently more reliable. Their speed is independent of the number of lines to be drawn since the paper moves at a constant rate under the nibs and through the toner bath. Unfortunately, they have to be roll-fed and the media have to be specially prepared with a non-conductive surface to prevent the charge leaking away before entering the toner bath. The choice of media is therefore limited and pre-printed sheets are not practicable. The picture on the paper is generated in the same way as a raster screen: by treating the paper as an array of little squares to be filled in. Plots therefore suffer from jagged curves and lines like raster displays. Multiple colour is not possible in most models currently available as it requires a separate toner bath for each colour and a separate pass under the nibs. There are models available with this facility but they are expensive for obvious reasons. A new design has

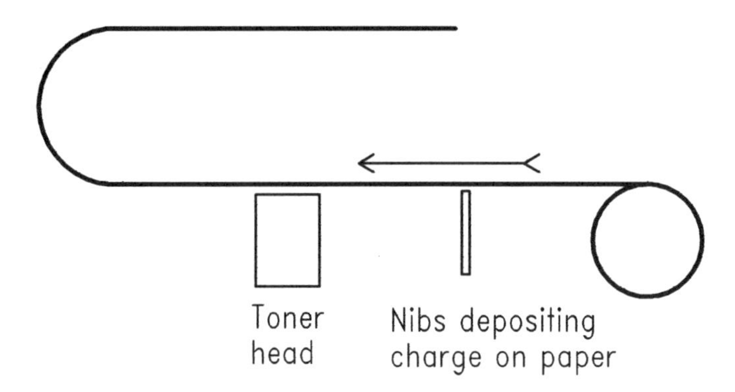

Figure 6.5 An electrostatic plotter

recently been launched which will make colour plots a reasonable proposition. The nibs are combined with the three toner baths in a single printing head which moves in a spiral path over paper wrapped round a drum.

INK-JET PLOTTERS

Ink-jet printing has been attempted on and off for about 20 years with little success but now the problems seem to have been solved for there are several ink-jet printers on the market. They operate with a single head which puts small dots down on the paper. They are relatively slow but have the ability to plot in colour. Their main use in CAD has been as screen copy devices for colour screens but may be used as the main output device in the future.

IMPACT DOT-MATRIX PLOTTERS

The dot-matrix printer has become the standard type for computer printing of medium quality. It is an impact printer which means that it operates by striking an inked ribbon against the paper. The striking is done by a vertical column of small needles in the printing head so as to produce a pattern of small dots the height of the character. The head moves across the paper laying down patterns of dots close to each other to form the line of characters. Since any picture can be produced by a pattern of dots the head can be programmed to print a drawing. However, due to the dot-forming process being mechanical the dots cannot be made small enough for a really high resolution drawing. The dot-matrix printer is therefore used as a screen copy device.

LASER PRINTERS

The superb quality and relative cheapness of these devices which can print graphics at high resolution means a good future for them. Models capable of handling the large widths of paper needed for engineering drawings have now come on the market.

COLOUR TRANSPARENCY PRINTERS

If you are making good use of colour-shaded pictures then you will need some way of making permanent copies of the pictures. Special devices for making colour transparencies from a high resolution colour graphics screen are available. As any photographer knows, colour transparencies are superior to prints or colour printers in capturing the brilliance of colour on a screen and they can be shown to a large audience. The devices intercept the electronic signals going to the screen and use them to drive a special internal display device.

OTHER SCREEN COPY DEVICES

Various other techniques have been used in screen copy devices such as the dry silver process based on paper coated with special photographic emulsion.

MAKING THE CHOICE

A number of decisions have to be made in specifying a plotter. The selection of models is wide. There is the choice between pen and electrostatic and the related decision on sheet or roll media: there is the decision on size and the difficult decision on speed.

1. Can you manage without colour and pre-printed sheets? If so, then you will be able to choose an electrostatic plotter and gain the benefit of its reliability and freedom from trouble in operation.

2. Can you manage without A0 drawings? Remember that text on a CAD drawing is legible down to 3 mm or less in height. If you can

do without A0 plots then you can save yourself money on capital and consumable costs by using an A1 plotter.

3. If you have chosen a pen plotter, do you really need instant plots? If you are prepared to wait until the end of the roll before getting your plots then you can save labour by using a roll-fed plotter and letting it run unattended.

4. Are you going to use your old dyeline printer or are you going to use the plotter for multiple copies? If you propose using the dyeline printer then you will need to use polyester film or vellum occasionally.

Having decided the specification, how do you make your choice? Avoid the enticement of features. At the time of writing there is a strange mismatch between plotters and the software which drives them. Plotter manufacturers seem to be selling on the basis of how many intelligent features they can put in. CAD systems, on the other hand, in order to be able to support a range of makes assume that the plotter is totally unintelligent. The world seems to be full of plotters which have features which will never be used! Furthermore, like the "intelligent" photocopier which feeds A3 paper when you want A4 copies the intelligence gets in the way! It seems to be very difficult to connect a new plotter to a CAD system. Plotters use fancy protocols for some reason instead of behaving like simple printers. You are advised to use a plotter which is known to work with the CAD software.

No plotter can work without paper, pens and ink or toner. Ensuring a good supply of good quality specimens of these consumable items will be one of your less interesting jobs! Poor quality consumables cause as much trouble as an unreliable plotter. Badly punched rolls jump off the sprockets, pens clog up, run out unexpectedly or will not take to the paper, and drawing ink flakes off the film. Deliveries are late or do not happen at all and so on. It is therefore a great comfort to have several sources of supply for your consumables and it is worth choosing a plotter which is widely used for that reason alone. Reliability is more important than anything else so look for machines which are soundly constructed and made by a manufacturer who looks as if he is going to stay in business, and keep the machine well maintained.

EXERCISE

As part of the process of getting new business, a company produces proposals which require drawings. The use of colour is considered an advantage and it has decided to employ a CAD system both for proposals and for engineering design. Discuss the options open in the choice of plotter if the company wishes to run the plotter with the minimum of attention during the day.

Chapter 7 The computer configuration

In this chapter we turn from the devices the computer controls to the computer itself. We can regard it as composed of the following parts:

- The central processor
- The main memory
- The disk memory
- The removable storage

We will give some idea of the contribution each part makes to the performance of the system and then discuss the difference between having a separate computer for each workstation and sharing one computer between several workstations.

THE CENTRAL PROCESSOR

The central processor (CPU) is the part which actually manipulates the data. It operates in a regular cycle and has a small number of fast registers of its own to hold temporary values. A typical cycle is:

1. Fetch one instruction from the memory.
2. Decode it to obtain:
   - The particular manipulation required.
   - The place in main memory where the data is stored.
3. Fetch the data from memory and put it into a CPU register.
4. Manipulate it as required.
5. Leave the result in a CPU register.

The first two steps of the cycle are the same for all operations but the last three steps vary according to the class of operation. The classes of operation are:

Memory reference As above. Sometimes step 3 is replaced by one in which the data in the CPU register is put back into the memory, in which case the remaining steps are omitted.

Operate Step 3 is omitted altogether. The contents of a CPU register is manipulated.

Control Data is not manipulated at all and memory is not accessed. Instead, the sequence of instructions is broken and restarted in another location in memory either unconditionally or depending on the result of a previous manipulation.

Input/Output Data is transferred between a CPU register and a peripheral device or a block data transfer is initiated between main memory and a peripheral device.

The CPU performs two classes of manipulations:

1. Mathematical

2. Logical, in which the contents of a register are treated like a pattern to be shifted from side to side, individual bits turned on or off etc.

As can be seen, one cycle operates on the contents of one memory register. The number of bits in a memory register is known as the *word length*. In large computers the word length is 32 bits, sufficient to represent large numbers with reasonable precision. In Personal Computers and minicomputers this is halved to 16 bits and in microcomputers even reduced to 8 bits in order to reduce the cost. The only way of handling numbers with reasonable precision in these cases is to combine several words to hold one number so that two or four cycles of the CPU are needed to process it. Clearly, the speed of the computer depends firstly on the speed with which the CPU cycles. This speed is usually quoted in *MIPS* or millions of operations per second. Another measure sometimes quoted is *MFLOPS* or millions of floating-point operations per second which is a measure of the speed with which the CPU does arithmetic whereas MIPS covers all the operations. But the speed of the computer does not just depend on the speed with which its CPU cycles. If the word length is half what it might be then twice as many cycles are required to do the same job, thus effectively halving the speed of the computer. This is what is behind references to 8-bit, 16-bit and 32-bit computers. They classify computers according to their performance.

The effect of word length is just one example of how performance can be degraded by the number of operations required to perform a particular task. Different computers have a different repertoire of operations (known as the *instruction set*). All computers have a sufficient repertoire to do anything needed of them but a particular task may need more operations to perform it in one computer than in another due to a lack of suitable operations in its repertoire. (In the same way that hand printing takes more operations than typing although the result is equivalent.) By now it should be clear that MIPS or MFLOPS are not a particularly good measure of the performance of a computer. The only way to compare performance is to give the two machines the same task to do and time them. In this way you are comparing the total system: hardware and software. It should be remembered that performance can be improved by carefully designed software and destroyed by badly designed software.

A particularly important influence on performance is the way arithmetic is done. For example, one can economise by taking advantage of the fact that all arithmetic can be done with just addition and subtraction. Multiplication and division instructions need not be provided; instead, the functions can be performed with software routines using repeated additions or subtractions. As a result the computer is slower. Where the multiplication and division instructions are not present they are sometimes offered in the form of an upgrade to be added to the CPU as a "maths co-processor" or a "hardware floating-point processor". The high calculation speed required to achieve a fast response makes this kind of upgrade common in CAD. As is described in "Encoding geometry" on page 75, most calculations are done using a special format for the numbers called *floating-point*. This requires more work to process so that floating-point instructions are often omitted from the repertoire of cheaper computers with consequent loss of performance.

A recent development relating to speed and instruction set is the discovery that, although a large repertoire of instructions looks desirable on the face of it, many of the instructions provided are not used very often and the act of providing the large repertoire introduces overheads which slow the CPU down. A faster machine is achievable by keeping the repertoire small in what are known as *reduced instruction set computers* or *RISC*.

Another technique for improving performance results from the fact that an operation is composed of a sequence of steps. The CPU can be speeded up if it is allowed to overlap instructions so that it is carrying out several instructions simultaneously. The overlap is not total. A typical sequence might be as shown in Table 7.1. Thus it deals with several instructions simultaneously, treating them like items moving along a production line. This is called *pipelining*.

Table 7.1 Pipelining

1	Instruction 1 step 1		
2	Instruction 1 step 2	Instruction 2 step 1	
3	Instruction 1 step 3	Instruction 2 step 2	Instruction 3 step 1
4	Instruction 1 step 4	Instruction 2 step 3	Instruction 3 step 2
5	Instruction 1 step 5	Instruction 2 step 4	Instruction 3 step 3

THE MAIN MEMORY

This is where the program and the data are kept when the program is running. It can be thought of as a great big array of numbered pigeon holes or cells. Each cell holds one word which can either be a number for processing or an instruction in the program. In each cycle of its operation the CPU fetches an instruction from the main memory, fetches a number from another cell in the main memory for processing or puts one back after processing. It is clear that the speed with which items can be fetched from or put into the main memory has a major impact on the speed of the computer. Usually, the length of the word in the memory cell is the same as that used in the CPU but sometimes they are different. You can be deceived by talk of a "32-bit" computer where only the CPU word length is 32 bits while the memory word length is 16 bits for the sake of economy. In these machines the CPU has to access the memory twice for each word it processes.

A technique used to speed up access to the main memory is the *cache memory* which is a small amount of extra fast memory used as a buffer between the main memory and the CPU. It is filled with the memory words which the CPU will be requiring next.

THE DISK MEMORY

The main memory holds the program and data while they are in use. The disk is a storage device which is considerably cheaper per unit of capacity than main memory although very much slower. It is a good place to keep programs and data in readiness for running since speed is not required, the large capacity making it possible to hold many programs. But the distinction between programs held ready for running and those actually running is not as sharp as you would think at first sight. Programmers, finding that their programs needed to be bigger than main memory, broke them up into sections which they held on disk. A root portion of the program in main memory then brought sections in from disk as needed. A logical extension of this was to relieve the programmer of the chore of organising this by providing facilities in the hardware and in the operating system. The programmer was allowed to write as if he had very much more main memory than he really had. The operating system kept the program on disk and brought into main memory whatever part of the program was needed when it was required. As the apparently large main memory was not real it was called *virtual memory*. CAD requires large arrays of data to represent the drawing so virtual memory is a valuable facility. It works well provided there is not too great a disparity between the real memory and the virtual memory, since whatever the programmer may see, sections of program still have to be read in from the relatively slow disk and this will slow up the program.

Disk stores consist of a metal disk coated with a magnetic recording medium. The disk rotates continuously at high speed and a dual-purpose reading and writing head positioned over it induces a pattern of magnetisation in the medium as the disk passes beneath it or, conversely, picks up the pattern of magnetisation previously induced in it. The head does not move while it is writing so that the data is written round a circular track at a particular radius. To access other data the head is positioned at another radius. The whole operation of accessing data on a disk involves sliding the head out to the required radius, waiting for the particular item to come round and then picking it up as it passes below. The time taken to perform the operation is known as the *access time*. Typical access times are measured in hundredths of a second which is slow compared with the millionths of a second required to access a cell in main memory.

To maximise the storage capacity the magnetic patterns need to be as fine as possible. The function of reading and writing uses magnetic fields which have the characteristic of broadening out with distance from their source. The result is that the head needs to be as close as possible to the moving disk. An ingenious method is used to achieve this. The air close to the disk surface is moving rapidly at the same speed as the disk itself. The head is given an aerodynamic profile like that of an aircraft wing which causes the moving air to generate a force sufficient to "fly" the head away from the surface. Since the region of moving air is very thin the head gets very close to the surface without actually touching it. The drawback of the method is that the gap is smaller than most particles of dust. Any particle getting in destroys the smooth air movement holding the head away from the surface and the head touches the magnetic medium and damages it. Known as a *disk crash* or *head crash* this usually destroys all the data on the disk. Careful measures have therefore to be taken to ensure that the air round the disk is extremely clean either by permanently sealing the disk and head in an enclosure during manufacture or by drawing in air through a fine filter while it is working. The former is the method used by the so-called "hard" disks or Winchester disks. A disk unit is shown in Figure 7.7.

THE REMOVABLE STORAGE

Anything held in main memory is volatile and disappears when the power goes off and the disk can destroy its data should it break down. The safest place for data is on some medium which does not need power to retain the data and which can be kept in a cupboard away from the device which writes to it. This is why no system is complete without some kind of storage device with removable media.

There are two quite distinct but equally important purposes served by removable media. The first is to guard against loss of data due to equipment failure, such as head crashes. The second purpose is the archival storage of data (in our case, engineering drawings) which is no longer being worked on but which will be needed some time in the distant future. High capacity for low cost is the principal requirement in both cases with stability over long periods of time being an important additional need for archive storage. The types of removable media storage devices currently in use are:

- Standard half-inch magnetic tape (Figure 7.1)
- Magnetic tape cartridge (Figure 7.2)
- Disk pack (Figure 7.3)
- Cartridge disk (Figure 7.4)
- Diskette (Figure 7.5)
- Optical disks

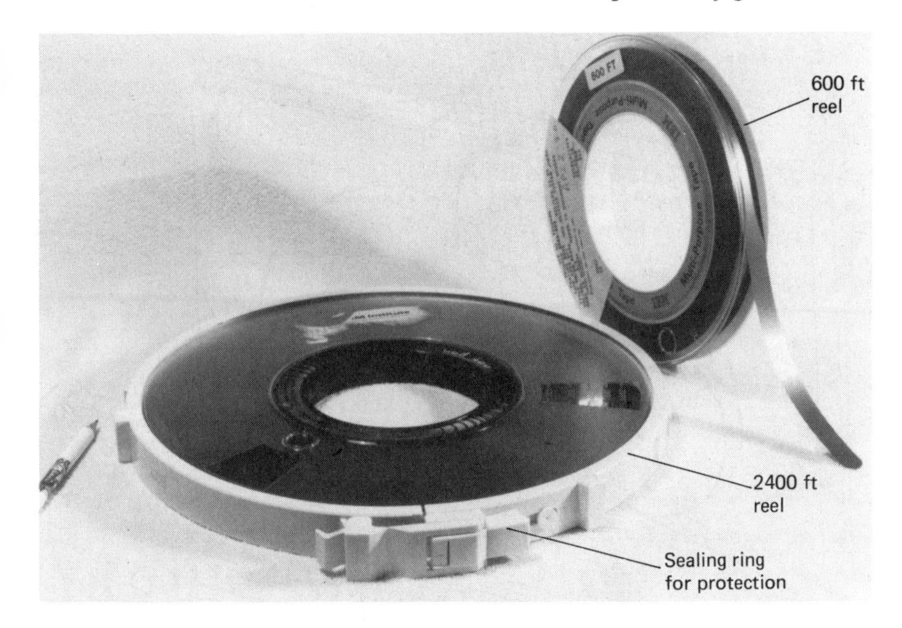

Figure 7.1 Standard half-inch magnetic tape

Magnetic tape units

Although expensive in equipment, the standard half-inch magnetic tape is undoubtedly the safest simply because it has been in use for so long and is standardised. It was in use as a storage device before disks were invented. The operation of a magnetic tape unit (MTU) (illustrated in Figure 7.6) is similar to that of its audio counterpart. The tape consists of a strip of mylar plastic half an inch wide and up to 2400 feet long wound on a reel. Powdered magnetic oxide held in a binding substance is coated on the plastic. The MTU draws the tape from the reel past a reading head and a writing head. A separate take-up reel winds up the tape when it leaves the heads. The writing head magnetises the oxide in a pattern representing the data as it passes. The reading head picks up the pattern of magnetisation as it passes. (The format is described in "Hardware data exchange standards" on page 203.) MTUs differ from audio tape recorders in that the data is often recorded or read in short bursts so that the mechanism drawing the tape past the heads must be able to accelerate and decelerate rapidly. To reduce the acceleration required of the two reels a loop of tape is maintained as a buffer either side of the heads. The initial movement only draws tape from the loop while the reel is coming up to speed at a slower rate. Sprung arms or vacuum chambers keep the loop in tension.

All your data is going to pass through your MTU and it must be relied upon not to corrupt the data. MTUs are complex electro-mechanical

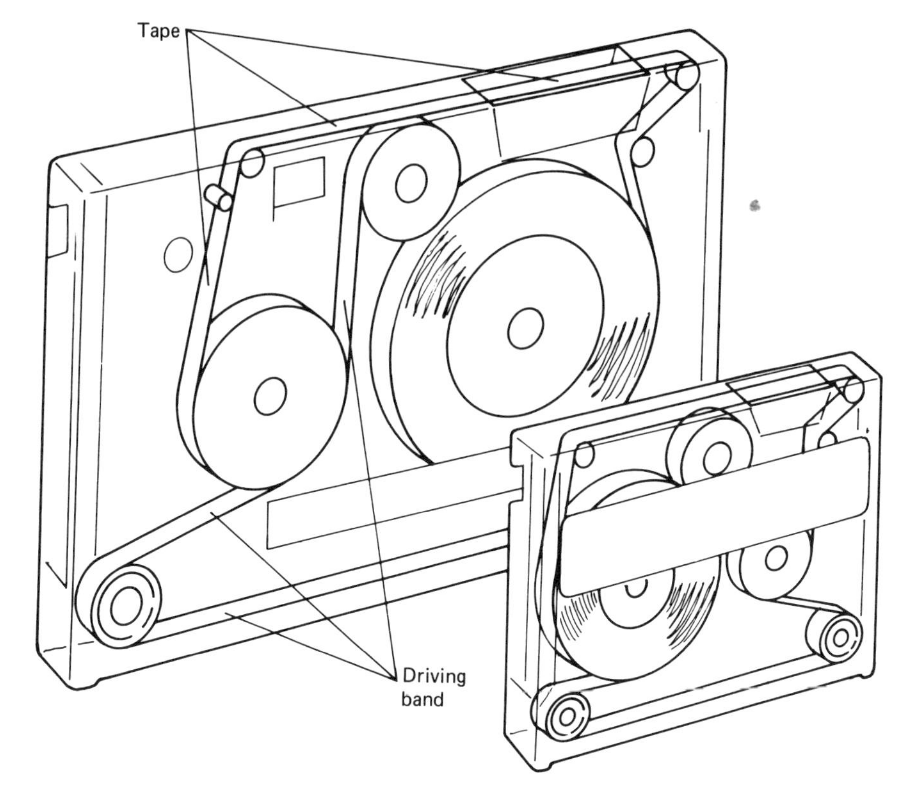

Figure 7.2 Magnetic tape cartridges

devices subject to wear and drift in adjustment. Regular maintenance is only omitted at the peril of losing expensive engineering designs. The reading head relies on extremely fine slots to pick up the magnetisation from the tape. The tape contains magnetic powder which rubs off and can clog the slot so that regular cleaning is important. The tape reel and the take-up spool have a large diameter but are driven at the centre. The acceleration force must be transmitted evenly from the centre to the periphery so that the reel behaves like a solid disk. This is only achieved by the drive mechanism winding the tape on to the reel with enough tension for the force to be transmitted from layer to layer in the tape by friction. If the tension is inadequate, sections of the reel stay bound together but in between one section and the next the single turn of tape joining them becomes pulled by the acceleration force and can be damaged. The process is known as *cinching*.

Much is known about the long-term stability of magnetic tape and it is by far the best means of archival storage. The precautions needed to take

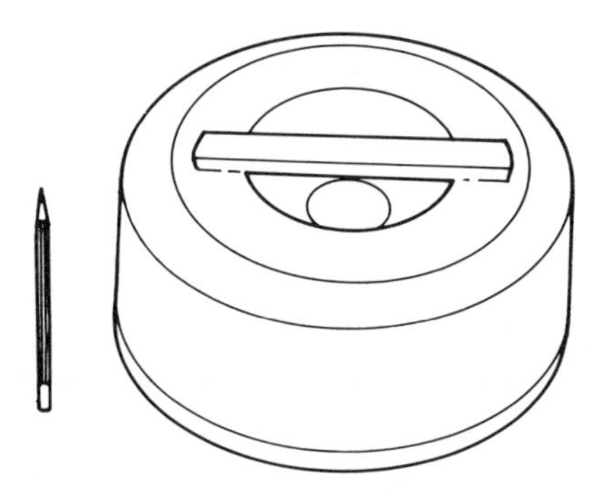

Figure 7.3 A disk pack

proper care of the medium are detailed in "Data security and contingency planning" on page 299.

Magnetic tape cartridges

The MTU uses separate reels. A number of cartridges have been developed and successfully used in which two reels are mounted in a single enclosure. One type is simply a high quality version of the ordinary audio cassette but other types have been developed where the design and construction are more robust and inherently reliable. The tape is ¼ inch wide. More recently the data processing industry has started to adopt a very robust high capacity tape cartridge using 1 inch wide tape. Its advantages over the traditional MTUs will mean that it will gradually replace them in due course.

Disk packs and cartridge disks

The disk pack is well established in the data processing business. It consists of a stack of disks on a common spindle which are lowered into the disk drive unit. It thus has all the speed of a fixed disk with the ability to remove the media for storage.

The cartridge disk is a cheaper version of the disk pack with smaller capacity.

Diskettes

Diskettes or "floppy disks" have become universal as storage media for Personal Computers. Rapid development of the technology under strong market forces has resulted in no less than seven different formats appearing in the space of about as many years, namely:

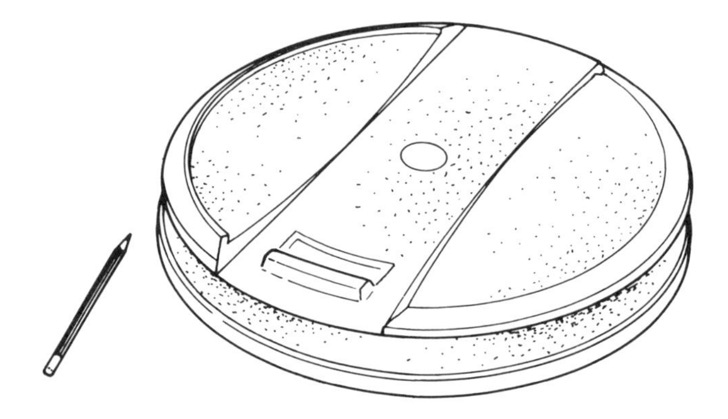

Figure 7.4 A cartridge disk

- 8 inch single sided
- 8 inch double sided
- 5¼ inch single sided
- 5¼ inch double sided
- 5¼ inch high density
- 3½ inch double sided
- 3½ inch high density

It would be advisable to wait for formats to become stable and standardised before attempting to use them for long-term storage.

Optical disks

Optical disks record data by changing the optical characteristics of microscopically small patches of a thin film deposited on a glass disk. The film is covered by a transparent plastic coating and the disk is written and read by a finely focussed laser beam. They offer capacities ten times more than the magnetic diskette. Early versions would not allow the data to be changed once written. Once again, the long-term stability has yet to be established. The film on which the recording takes place is chemically reactive and the plastic coating is not entirely impervious to water vapour from the atmosphere.

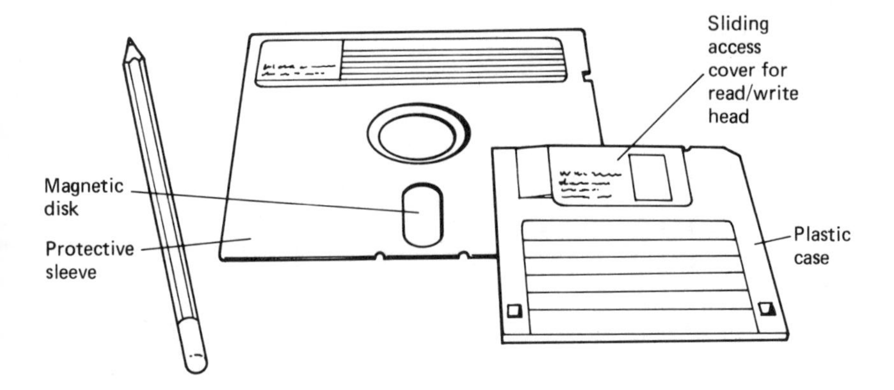

Figure 7.5 Diskettes

CENTRALISED AND DISTRIBUTED SYSTEMS

Twenty-five years ago a computer was a major capital investment which had to be used as efficiently as possible. It was therefore a centralised resource. Techniques in the operating system known as *multi-tasking* or *time-sharing* allowed many users to use it simultaneously in an interactive fashion. Since then the cost has dropped steadily in a trend which is continuing. First to be developed was the 12-bit and then the 16-bit minicomputer which handled simple programs one at a time and was within the scope of the large department or small company. The principal use was in the control of machinery. The minicomputer was then extended to the 32-bit super-minicomputer which was able to do multi-tasking or time-sharing. It was at this point that CAD systems were introduced as the economics were right for a single super-minicomputer supporting up to six graphics terminals in a CAD system. Minicomputers achieved cost reduction by the use of micro-electronics. The development of complete CPUs on a single micro-circuit led, at about the same time as the arrival of the super-minicomputer, to the 8-bit microcomputer. These had a similar performance to the early minicomputers. Further development of micro-electronics produced 16-bit microcomputers on which current Personal Computers are based which have a similar performance to the later minicomputers. Time-sharing was no longer needed on account of the low cost. With the addition of a maths co-processor and a high resolution graphics screen the Personal Computer is capable of providing a reasonable range of CAD functions in a completely self-contained workstation. The production of 32-bit CPUs on micro-circuits now provides enough performance at a low enough cost for a full range of CAD functions on a single workstation.

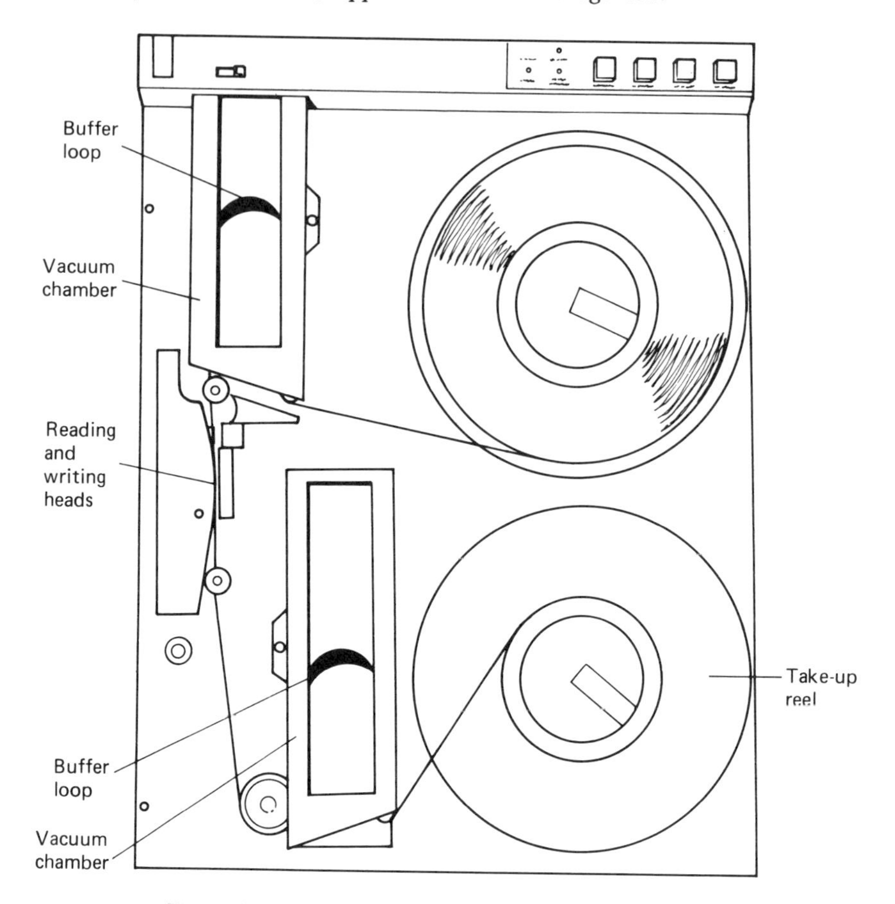

Figure 7.6 A magnetic tape unit

All the types of computer described currently coexist and all have CAD software available although an 8-bit microcomputer will not have the performance to make CAD worthwhile. A purchaser of a CAD system therefore currently has a choice of CAD on any of the following:

1. A single centralised shared computer or "mainframe"

2. A single super-minicomputer supporting a cluster of workstations

3. A set of self-contained 32-bit workstations

4. A set of Personal Computers

The advantages of each workstation having a CPU to itself are clear: the response time will be consistent and not drop off when other operators happen to be demanding heavy computation, and the installation can be

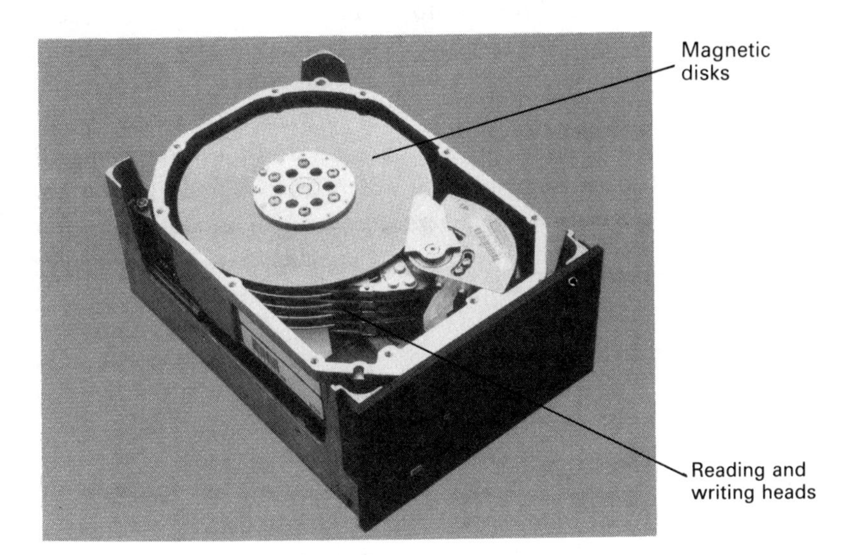

Figure 7.7 A disk memory unit

extended indefinitely in easy steps by the purchase of additional workstations. The principal reason for shared CPUs is economic and this will disappear as the cost of micro-electronics continues to decrease. Future CAD systems will therefore be based increasingly on self-contained workstations.

Essential shared facilities

Even in a distributed system there are certain facilities which for logistical reasons have to be centralised and shared. These are:

File storage

The CAD drawings must, by their nature, be controlled centrally. The situation where two people work on copies of the same drawing simultaneously must be prohibited. Checking and issue must be controlled by a single authority responsible for the work of the department.

Data security procedures are also best done centrally to ensure that they are performed in a timely and proper manner. One of the few disadvantages of distributed systems is that the chore of backing up (see "Back-ups" on page 300) has to be done at each workstation but unless the right steps are taken nobody feels sufficiently responsible to perform the procedure. Such a state of affairs can continue until vital drawings are destroyed by disk failure.

A safer way of working is to arrange that drawings can only be stored centrally in a *file server* machine which controls and logs the work being done on them. Back-up procedures are then performed under the supervision of the manager at the file server.

Plotting

Plotting out is naturally a centralised shared function as there is no need for a plotter at every workstation.

Drawing issue

The issue of drawings in machine-readable form to other departments or other companies and the extraction of parts lists, NC data etc involves some kind of centralised control and possibly some kind of conversion operation. Both of these require a centralised shared machine.

A potential advantage of distributed systems is tolerance to hardware failure. If there were no shared facilities at all with all the work being done on a set of workstations, a single hardware failure can only remove one workstation which is only part of the total capacity. Shared facilities modify the case as a failure in a central facility could stop all the work. To avoid this the system of hardware, software and administrative procedures should be designed so that work can continue for a limited period without the shared equipment. The file storage is the only one which requires thought. One is tempted to do all storage in a file server and provide no local storage for drawings in the workstation at all. Failure of the file server would then cause all work to stop. A better scheme is to provide enough disk locally for temporary storage of just the drawings currently being worked on. Drawings are then read from the file server at the start of the session on the workstation and saved to it at the end. Software would be needed to ensure that the work was indeed saved at the end of the session but that would be relatively easy to provide. Failure of the file server in this scheme would not hold up work until the start of the next session at a workstation.

THE FUTURE OF DISTRIBUTED SYSTEMS

There is a number of facilities which are feasible in a distributed system but which are not generally available at the moment. They consume extra resources in the operating system and require additional knowledge and decisions to be made when setting up the system. As system purchasers and managers become more knowledgeable, a market will be created and these features will start to appear.

Partitioned data

The current structure of networked systems is simply a central file server supplying storage to all computers on the network. There is no logical reason why the data should not be distributed as well with the physical location of files being chosen by the system designer according to the way in which they are going to be used. For example, a file heavily used by one computer but only lightly used by others is best located on that computer, or a file which is the sole responsibility of a particular department may be best located on the computer managed by that department.

Transparency

Although of concern to the system manager the exact physical location of data is of no consequence to the user. Therefore, in reading in a file on his workstation he should not have to remember where it is located, whether on his workstation, the file server or even another workstation (if that is the way storage happens to be set up). There should be a single file management structure for the whole distributed system in which the user requests files by descriptors meaningful to him, related to the way his work is structured (by project, activity etc). Implementing this requires that the location of files be known to the file management software and that the networking programs be embedded in the file management software.

Replicated data

If identical copies of files are held on two separate physical devices, a very robust system results. Both devices would have to fail simultaneously to lose the data or to lose access to the data. A distributed system provides the means of connecting separate devices together but a further facility has to be provided by the file management system for replication to become a practical proposition. It must ensure that any alteration made by the user takes place in both copies. Not only must it do this under normal working conditions but also where failure has made one copy inaccessible and where failure takes place while it is in the process of bringing one copy into line with the other.

Data security

With a unified structure for all the data wherever it is distributed, back-ups for the whole data can be performed by any selected computer in the system so that files do not have to be located on a central file server to ensure that they are backed up. Where it is replicated, data can be recovered automatically from another copy in the event of the disk failing. Thus, with the right software, a distributed system can become very reliable and robust.

THE EVOLUTION OF THE COMPUTER CONFIGURATION

In summary we can see a distinct trend towards networked workstations. At first, the workstations will use a central file server which itself will acquire a full database management system to control the CAD drawings and then all related documentation. The database will become linked to the manufacturing database and become part of a corporate database in a fully integrated environment. Support for fully distributed data will be incorporated into the operating system and the file server will then become the device which performs back-ups, input and output, and links to other departments and the company databases. The data will be distributed among workstations according to the function it serves with a fair amount of replication for the sake of robustness.

EXERCISE

List the factors which affect performance and the factors which affect reliability.

Chapter 8 Encoding geometry

Before considering the large range of facilities provided by CAD software we will look at the way the computer represents the drawing in its memory. We can divide the information in a design into two types: numerical data specifying lengths etc and qualitative information such as notes, identifiers and descriptors of various kinds. All information has to be held in the memory in the form of a list of integers for that is how the memory is organised. As mentioned in "An introduction to computers" on page 5, anything can be represented by a number. All that is needed is a conversion table to convert between the number and the description the human being uses. The software therefore codes all qualitative information into lists of numbers using whatever conversion tables the programmer might invent for his own convenience. Most of it will be kept as text for which there is an international standard conversion code.

ENCODING DISTANCE

Strangely, it is representing the numerical information which requires some thought. As mentioned in the previous chapter, each memory storage location holds a group of binary digits (bits) known as a word, the number of bits in the word being fixed for a particular model. This means that a location cannot hold a number with more than so many digits. By way of illustration let us assume that the location holds up to seven decimal digits (and that the computer uses decimal notation instead of binary). As we need to handle numbers with fractional parts we put an imaginary decimal point two digits in from the right leaving five digits for the integer part as follows:

```
1 2 3 4 5 1 2
         .
```

In placing the decimal point there we limit the resolution to 0.01. We cannot add or subtract less than 0.01 and we can only represent real-world dimensions to the nearest 0.01. Although it is satisfactory for typical lengths of around 50 mm, small distances of, say, 0.05 will have a percentage accuracy of only plus or minus 10%. The resolution as an absolute quantity is constant at 0.01 but when calculated as a percentage of the main value it increases as the main value decreases. This may not be serious in engineering where the accuracy of machining is an absolute value rather than a percentage but it does mean that the programmer would have to provide enough decimal places to satisfy the highest precision application likely to be encountered. Now suppose the number is 99999.99 and we add 0.01. There are no more places left and the program has to stop. This is known as an *arithmetic overflow* and is catastrophic. The only way it can be avoided is to provide enough places for the largest number likely to be used.

Suppose that the programmer, bearing these things in mind, provides for a maximum of 9999 metres to cater for civil engineering. At the same time he provides for a maximum accuracy of 0.01 mm to cater for precision engineering. The numbers will have the format of 9999999.99 using nine decimal digits. Every number will occupy this space in the memory however small it is yet the first five digits would always be zero in normal engineering and the last three digits, representing 2 mm, would be meaningless in civil engineering distances of 9 km. The amount of space used up in storing numbers affects the performance of the system with respect to the amount of storage required by the drawing and the speed with which the program can find an entity, since more memory has to be traversed.

Where numbers can have a very large range of values the accuracy tends to vary with the size of the number, the percentage accuracy staying roughly constant. In scientific work, percentage accuracy has more meaning than absolute accuracy. Because of all this, a way of representing numbers was designed which makes efficient use of the storage space over a very wide range of values. It is known as floating-point format. To convert a number into floating-point format the decimal point is moved left or right until it is just in front of the first (non-zero) digit and the number of positions it was moved is noted as a positive or negative integer. The original number is thus broken into two parts: a fraction between 0.1 and 0.9 and a signed integer. Digits are trimmed from the right-hand end of the fraction to fit into whatever space has been allocated. Thus, 1234567.89 becomes 0.12345 and 7, assuming that five digits is the space allocated. This is sometimes written 0.12345E7. Space has to be allocated for the integer but even a small space permits an enormous range of values. One digit and a sign in this decimal example will allow numbers in the range 0.0000000001 to 100,000,000 to be used. Trimming the right-hand side of the fraction means that the percentage accuracy is about the same whatever the size of the number. It also has an important effect on the arithmetic. Consider the following where 0.01 is added twice to 999.98:

```
 999.98 + 0.01 =  999.99 = 0.99999E3
 999.99 + 0.01 = 1000.00 = 0.10000E4
1000.00 + 0.01 = 1000.01 = 0.10000E4
```

The value 0.01 makes a change to 999.99 but no change to 1000.00. Because of this the criterion for equality between numbers is not hard and fast and a tolerance has to be set as the difference below which two numbers are considered equal. This has implications in CAD when the program has to decide if two things are coincident or not. In some systems the user is not made aware of the tolerance used. In any case, the precision of the system is usually set so that it does not matter to him.

The characteristics of floating-point numbers contrast with those of fixed-point numbers. The latter have a limited range dependent on the number of digits allowed, a fixed absolute resolution but a variable percentage resolution. Floating-point numbers have a very large range, a constant percentage resolution dependent on the number of digits allocated and an absolute resolution which varies over a wide range. It follows that, since floating-point numbers are used in CAD/CAM software, the percentage resolution is the factor most likely to suffer if the programmer chooses to be mean on storage. The range will always be impressively large.

ENCODING POSITION

To encode position the software uses the well established mathematical notation of Cartesian coordinates. Two axes perpendicular to each other are set up and the position of a point is determined by its distances (known as coordinates) from each axis. In a three-dimensional (3D) system, a third axis perpendicular to the others is used and a point will then have three coordinates. The point where the axes intersect is known as the origin. Each entity in the drawing will be positioned and defined by one or more points, each consisting of a set of coordinates. Lines will have end points, arcs will have centres and end points etc. As the designer specifies the various constructions the software calculates the coordinates of the new points created using coordinate geometry. In all this the exact location of the origin will not matter as all positions in a drawing are relative. Sometimes the designer needs to input data in terms of coordinates rather than in terms of geometric constructions. In this case, he will need an origin for the coordinate system he is using and the software will allow him to specify a temporary local origin or even a complete set of axes rotated from the usual ones. But this is only an input facility. The program will convert the data into its own coordinate system whatever that is.

SCALE

Newcomers to CAD/CAM usually find scale confusing because there is more than one scaling factor in operation at a time. In distinction to paper drawings, CAD/CAM systems record distances in real-world units and numbers no matter how big the item being designed. They can do this simply because the design is encoded in numbers. A kilometre in the design is a kilometre in the CAD drawing, unlike a paper drawing which has to apply a scale because it cannot hold a line one kilometre long. The user of CAD/CAM can therefore input distances as they actually are in the real world, which is a great convenience particularly when importing a portion of one drawing into another.

Drawing scale

Eventually, the one kilometre long line has to be plotted out on paper. At this point, a scale has to be applied as it does with paper drawings in order to get it on the paper. The user is asked to decide the scale and to make the drawing easy to interpret he makes it a convenient round number as he would in the case of paper. It therefore serves the same function as the scale of paper drawing practice. If the drawing contains several views, each one may need a different scale so there may be several plot scales for the different views as in paper drawing.

Display scale

Besides plotting the drawing out on paper the program has to display it on the screen. Here the scale will be whatever makes the chosen portion of the drawing fill the screen. Where there are several views at different scales the individual scale for each view will be applied first, before applying the scale for the screen, in order to maintain the appearance of the drawing when plotted out. The user will be continually changing the display scale according to the amount of detail he needs to see. Where there are several windows on the screen, each one will be treated like a separate screen with its own scale.

Input scale

All the previous scales are concerned with output and display. But sometimes the user may wish to enter distances in different units to those in which the drawing is held. Some programs allow the user to set a unit of measure of his choice as a convenience in inputting data. This, of course, is yet another scale, but applied only during input. The various scales are shown in Figure 8.1.

ARRAYS AND MATRICES

Considerable use is made of tables and lists of numbers in the data describing a CAD drawing. Even such a basic thing as a point requires a list of at least two numbers: its coordinates. A line drawn from the origin to the point has a direction and a length so that the pair of coordinates defines a vector. (A list of numbers is also called a vector.) Any straight line can be defined by the two points at its ends: two lists of two numbers each. A natural way of writing down the two lists is as a table of two columns, each with two numbers. A table of numbers in rows and columns is known as an *array*. Matrix algebra, the field of mathematics devoted to manipulating such arrays (referred to as *matrices*), provides a powerful way of solving the equations encountered in the coordinate geometry of a CAD system and in carrying out other calculations. Computers provide special facilities for rapidly accessing individual numbers in an array because of the extensive use made of arrays in mathematics.

The user is not normally made aware of the use of matrices in the CAD software but one particular application of interest is the transformation matrix. Three very common operations performed on a part of a CAD drawing are moving it, scaling (magnifying or reducing) it and rotating it. It is possible to represent any of these operations (known as *transformations*) with a 4 x 4 matrix (known as a *transformation matrix*). The alteration to the geometry is carried out by a matrix multiplication between the transformation matrix and the vectors representing the points involved, the result being a new set of vectors representing the new positions of the points. Performing a combination of scaling, moving and rotating is a matter of multiplying three matrices together with the points involved. What is more, the transformation matrices can be multiplied together on their own before being applied to any points. This produces a single transformation matrix defining a particular complex transformation. Any particular combination of movement, scaling and rotation can thus be represented by a single 4 x 4 matrix. The technique is elegant and its application can sometimes be seen in the way a CAD system handles transformations. A fuller treatment of transformation matrices can be found in Reference (20).

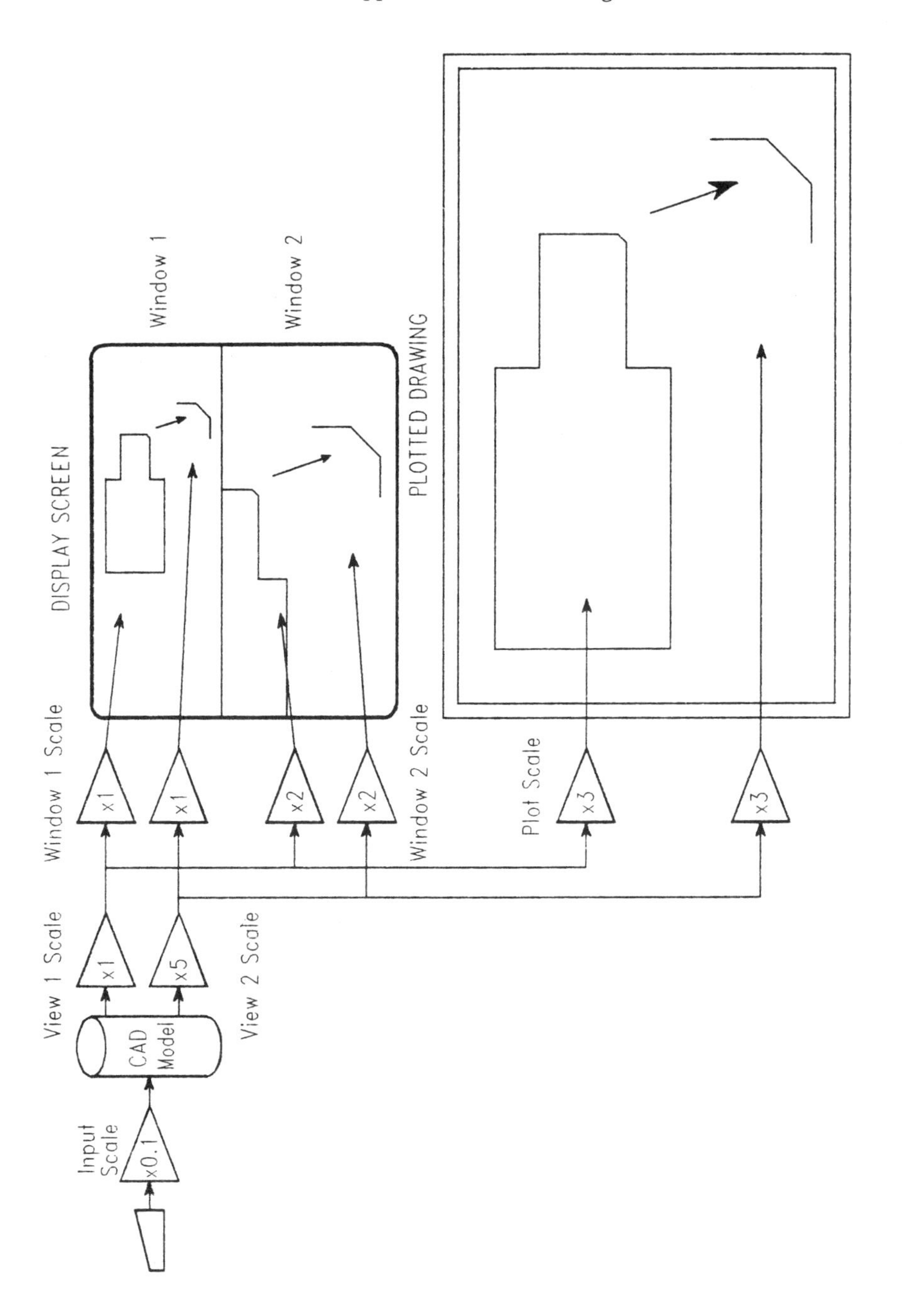

Figure 8.1 Scaling factors in a CAD system

LOGICAL LINKS AND DATA STRUCTURES

The previous section introduced the idea of collections of numbers arranged in a particular way. To put it more generally, the arrangement in rows and columns expressed the particular relationship the numbers had to each other. A computer memory is structured in the form of a sequential list of numbers. It represents the array in its memory by going along the rows in turn. The links are thus preserved by the sequence in which the numbers are held.

A rectangular array is quite a simple arrangement. Representing the complexity of real objects requires more complicated links. Consider the three-dimensional framework of lines defining the edges of a rectangular block shown in Figure 8.2. The only numerical parts of the framework are the coordinates of the eight points at the vertices. The rest consists of lines joining particular pairs of points and all that is needed to define a line is to name the two points it joins. The data defining the framework therefore consists of, firstly, a table of eight points as shown in Table 8.1. Secondly, we have a table as shown in Table 8.2 defining the lines which simply links points together - it contains no coordinates at all. Such a table represents the topology of the object. If the object was a solid, there would be faces as well, which could be represented by a table as shown in Table 8.3 defining the edges bounding each face.

This example illustrates the use of links to hold topology. Links are used for other purposes such as connecting the dimension in a drawing to the thing dimensioned or attaching descriptive information (part number, finish etc) to an object. It can be seen that with all this cross-linking the structure of the data in a CAD drawing can become quite complex. References (20) and (35) deal with data structures in CAD in further detail.

EXERCISES

1. A particular system claims to handle a wide range of distances. What else affects the accuracy of the CAD data?

2. Give both the hardware and software reasons for the fact that, unlike a paper drawing, the lines on a CAD screen do not get thicker when it is zoomed.

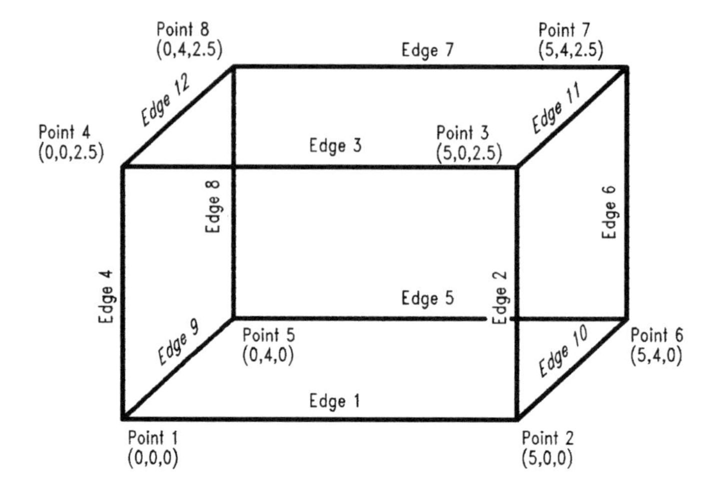

Figure 8.2 Geometry and topology

Table 8.1 Coordinates of points of rectangular block

Point	**Coordinates**		
	X	***Y***	***Z***
1	0.0	0.0	0.0
2	5.0	0.0	0.0
3	5.0	0.0	2.5
4	0.0	0.0	2.5
5	0.0	4.0	0.0
6	5.0	4.0	0.0
7	5.0	4.0	2.5
8	0.0	4.0	2.5

Table 8.2 Lines linking points of rectangular block together

Line number	First point	Second point
1	1	2
2	2	3
3	3	4
4	4	1
5	5	6
6	6	7
7	7	8
8	8	5
9	1	5
10	2	6
11	3	7
12	4	8

Table 8.3 Edges bounding the faces of an object

Face	First edge	Second edge	Third edge	Fourth edge
1	1	2	3	4
2	1	10	5	9
3	5	6	7	8
4	7	12	3	11
5	12	4	9	8
6	2	10	6	11

Chapter 9 The repertoire of graphical entities

With this chapter we embark on a detailed discussion of the bewildering array of facilities provided by the software. The principal aim is to remove the bewilderment and to help form some idea of the relative value of each facility.

The foundation for the facilities is the way the software stores the design in the memory of the computer. There are many ways in which it could organise the data representing the design but in order to make it easy for the human user to communicate with the software the structure of the data must model what the user is familiar with, behaving in a way which is predictable from his point of view. Thus, since the user is a designer, he should not have to specify alterations as changes to algebraic variables but instead as changes to lines and circles or other geometric objects. All CAD systems model the design as a collection of items (*entities*) which are familiar geometric objects for the most part. In addition, certain items are included to organise the design by grouping entities together and to provide the descriptive information previously given in notes. The various proprietary systems differ in the precise set of entities employed and in the properties allowed them. The remainder of the chapter will review the different entities used.

TWO- OR THREE-DIMENSIONAL GEOMETRY

One of the first things you notice about CAD systems is the division into two-dimensional (2D) and three-dimensional (3D) systems. 3D CAD software allows the designer to do something which is impossible with paper: to lift the pencil off the paper, going in whichever direction he chooses in space and still trace out lines. Lines going in all directions in space are rather like a piece of crumpled wire netting so 3D systems are said to hold a *wire-frame* model of the design as shown in Figure 9.1. Although it is nearer reality than a drawing on paper by virtue of the third dimension, it is still only a line drawing and lacks all the other properties of a solid object such as surfaces and a knowledge of what is inside it and what is outside it. 2D systems, on the other hand, just model the lines on a piece of paper. Both styles of system are in successful use and the choice between them is a matter of what the system is going to be used for, the amount the user is prepared to pay for it and personal preference.

POINTS

Points are fundamental to all CAD software. Lines are defined by their end points, texts by a string of characters starting at a particular point, and curves are constrained by various points along them. In the data structure representing the drawing in the memory, each point is held as a pair of numbers which are the two coordinates in a Cartesian coordinate system. Every drawing therefore has an origin for its coordinate system, usually but not always the extreme lower left-hand corner. Although points are fundamental, not all systems provide points as entities which can exist on their own. Instead, positions are marked by the ends of lines or the intersections between lines.

STRAIGHT LINES

Of course, a CAD system would not be a CAD system without straight lines in its repertoire but there is more than one way of handling straight lines. The basic building block of engineering drawings is not so much the isolated straight line as the closed outline representing a physical object. Although geometrically it is a collection of separate straight lines with the starting point of each one coincident with the end point of the previous one, it needs to be manipulated as a single unit. Two different approaches are possible. The software can either let the user create each line separately or make the user create the lines in sequence as an explicit composite line. The first method is easy for the user to learn but since he can draw lines which do not join up or which branch, there is no way for the software to be sure what the closed outline is in all cases. When it needs an unambiguous closed outline for an area calculation or solid modelling, for example, it has to engage in a dialogue with him to establish which particular set of lines is to be used. In the second method the software can be sure from the outset that it has a closed outline without having to ask the user.

Sometimes both types of line are made available. This is unnecessary since an isolated straight line can be represented by a closed outline composed of two segments lying on top of each other. In one system, a single data structure is used for any closed sequence of straight line segments and arcs. Editing the lines requires some training but at least there is only one kind of line to understand.

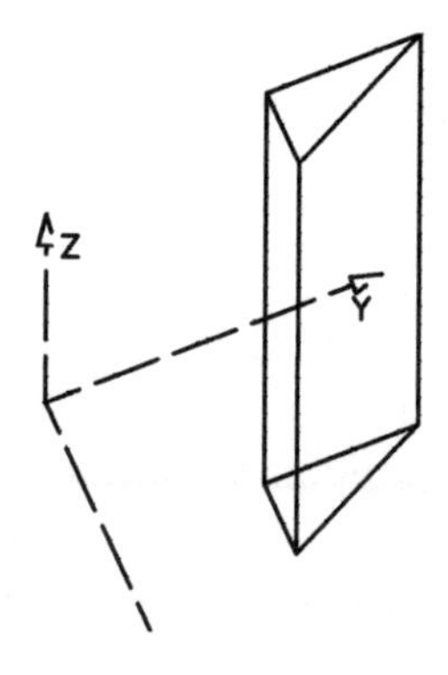

Figure 9.1 A 3D wire-frame model

CIRCLES AND ARCS

Like straight lines, circles and arcs are fundamental to CAD. Some systems provide circles and arcs as two distinct entities although this is unnecessary since, by treating a circle as a 360° arc, the same data structure can hold either. One can construct an arc in many ways and a good system will provide a comprehensive range of facilities. There may be an advantage in having different data structures, and hence different entities, specifying arcs in different ways in order to have them behave in different ways during alterations. For instance, holes need to be held in place by their centres and specified by their radii or diameters whereas fillets are held in place by the adjacent line segments and not fixed to particular centres.

TEXTS

Texts are again essential. Although they do not embody any geometry belonging to the design they have geometric data in them to position them in the drawing. A text entity will consist of at least the first five items of the following:

1. The string of characters

2. A point to locate the text

3. The height of the characters

4. The orientation of the text line

5. Justification

6. Font or style

7. Slant

8. Spacing

9. Aspect ratio

The justification is the position of the line of characters in relation to the point which fixes its location. It may have its left end, its right end or its centre on the point. In addition, it may be possible to locate the bottom of the line or the top of the line or even the middle of the line on the point. If a number of different styles of text have been made available then the particular style will need to be specified. Alternatively, or additionally, it may be possible to tilt the characters over by a specified amount in order to achieve an italic effect. Another way of getting various styles is to alter the spacing between letters or the ratio of the height to the width of the letters.

When text is plotted out or displayed on the screen it is generated like the rest of the drawing with strokes of the pen. In order to conserve storage space the line segments making up the characters are not usually stored in the drawing as such. Instead, the data described above is stored and a special routine turns the data into the appropriate line segments during the plotting or screen drawing operation.

CONIC SECTIONS

Of course, a circle is one particular conic section but some CAD programs include the remainder of the set - ellipses, parabolas and hyperbolas - in their repertoire. By doing so they provide for those occasions when a hole has to be shown viewed obliquely, for a wider range of curves and for transformations which stretch circles more in one direction than another.

DIMENSIONS

A dimension is a collection of lines, arrows and texts. Some CAD programs simply put dimensions in the drawing data as such, possibly binding them into a group, but others actually distinguish dimensions as separate entities. The advantage of doing so occurs if the program recalculates and redraws dimensions whenever the geometry they describe is altered.

SPLINES

Splines are like French Curves but being mathematically precise and very controllable, they more than replace them. They provide the means of creating smooth curves with a wide range of shapes. An important characteristic is that the shape is defined by just a few control points. Depending on the mathematical techniques used by the software the control points are used in a number of ways. Two of the control points will usually fix the start and end of the spline. The curve may then be forced to pass through or near the others or be tangent to lines joining pairs of them. The control points are thus means of defining positions through which the curve must go or the direction it must follow at particular places. This should be distinguished from the technique of defining curves in which a large number of points is laid down and the curve fitted to them. Considerable design power is provided by the ability to determine independently what direction the curve should take at any position. This is illustrated in Figure 9.2 where three radically different splines have been produced through the same three points.

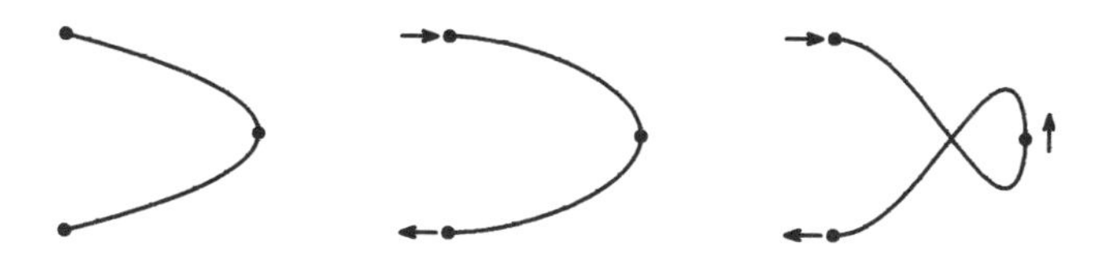

Figure 9.2 Controlling a spline curve

The particular spline generator used constrains the curve to pass through each point. The user may apply an additional direction constraint at each point. The three different splines have been made by simply altering the direction constraints. The ability to apply direction constraints is important for joining two curves together. If the directions of the curves can be made the same at the point where they join the transition from one to the other is smooth. The direction is continuous across the join. The rate of change of direction, or curvature, may not be continuous across the join but this often does not matter. In some cases, where the splines are to be used as the basis of surfaces used for styling, it may be necessary to make the curvatures continuous as well, since reflections from a surface show the curvature.

For an account of the underlying mathematics of splines see References (20), (34), (35) and (37). Curves are handled differently in CAD to the familiar graphs of functions in coordinate geometry. Instead of deriving the x, y and z coordinates from equations relating them as, for instance, in the equation for a circle:

$$x^2 + y^2 = r^2$$

a separate variable is used which, roughly speaking, represents the position along the curve. A value is chosen for this variable and then each coordinate is calculated from it. In the case of the circle the variable could be the angle of a radius, a. Each coordinate is then:

$$x = r\cos(a)$$
$$y = r\sin(a)$$

The method has a number of advantages, one of them being that there is only one solution to the equations where there might be several using the other way. In the case of the circle the solution produces two values of y for every one of x. Curves defined in this way are known as *parametric curves*.

A term that is often used in this connection is "NURBS". This is an abbreviation for "Non-uniform rational B-spline". The NURBS has become popular with CAD software writers recently. It represents a particular mathematical technique whose value must be judged (as with other techniques) on the extent to which it offers the user more accurate or more efficiently produced curves.

SURFACES

Surfaces should not be confused with solids although they often look the same on the screen. A surface has no thickness or no "inside". The nearest it can get to a solid is to enclose a volume. But it can exist without enclosing a volume (it is difficult to make it enclose a volume without leaving a hole somewhere) whereas a solid cannot avoid enclosing a volume. They are provided because certain industries (automobile and aerospace) need to reproduce complex curved surfaces precisely and CAD/CAM provides the only way outside of workshop techniques to design and specify them.

Surfaces are generated in most cases by moving a line about in space just as a line is generated by moving a point. There are two levels of complexity. The simpler type of surface is generated by moving a straight line about and they are often called *ruled surfaces*. A ruled surface which has been produced by sliding a line along a curved rail is shown in Figure 9.3. With these, there is always some direction on the surface along which a straight line can be laid. Examples of ruled surfaces are cylinders and cones. More general, and therefore more complex than a ruled surface, is the doubly curved or sculptured surface which is generated by moving a curved line about in space. Sometimes the curved line is made to change its shape as it moves about. An example of a doubly curved surface can be seen in Figure 4.1 on page 32.

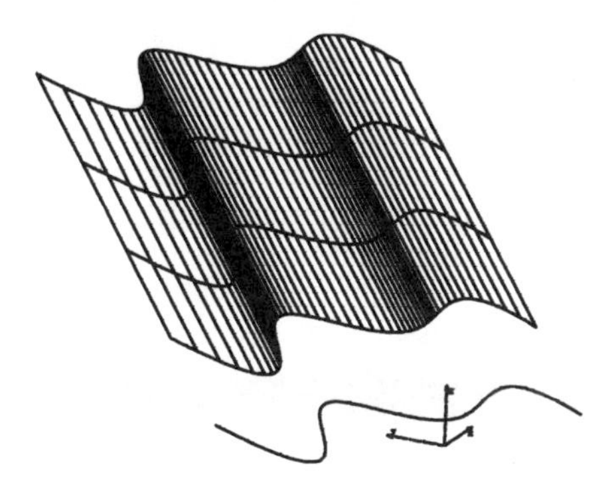

Figure 9.3 A ruled surface

SOLIDS

The characteristic feature of a solid, as suggested in the previous section, is that it is bounded by one or more closed surfaces. Solids are nearly always generated by moving closed lines about in three-dimensional space. A solid is the only entity which fully models real objects. It is an unambiguous definition of a closed space whose volume and other properties can be calculated without further definition or resolution of ambiguities by the designer. While surfaces can provide many of the functions of solids, such as volume calculation and even the visualisation of shape through colour-shaded perspective projections, their limitations become apparent when objects with voids inside them have to be modelled.

The close parallels with real engineering components are exploited in some of the generation and editing facilities provided which mimic commonly used machining operations.

PLANES

Planes have a number of uses. A line is defined by the intersection of two planes and a point by the intersection of a line with a plane. 3D drawing can be assisted by temporarily limiting the geometry to that lying on a selected plane. A plane can define a section through a solid and produce a curve from a section through a surface. Surfaces can be defined in terms of tangency to one or more planes. Besides planes as explicit entities, the view in a window involves an implicit plane which is sometimes used in, for example, confining new geometry to a particular distance from the view plane.

INVISIBLE NON-GEOMETRIC ENTITIES

There is a number of entities used in the drawing data which are not geometric and are invisible. These are groups, symbols and attributes. The invisible entities are not possible at all in a paper-based design system. They provide facilities not available before. In particular, they make it possible for the CAD drawing to be a complete specification and statement of the design.

Groups and symbols

A group is a collection of entities bound together. The binding bears no relation to the geometry and is completely invisible in the drawing although in practice the entities are usually near each other. The function of a group is to hold together all the entities which belong to the same physical object: the lines making up its outline, text describing it, text listing its component parts etc. Two useful facilities result. All the items can be moved about the drawing together in a single operation without having to put a box round them and a parts listing program can find all the components of an assembly unambiguously.

If the software permits groups to be included in a group then the drawing can represent (invisibly) a complete family tree of assemblies, sub-assemblies and components for a parts listing program to extract. It becomes more than just a definition of the geometry of the design.

Closely related to groups are symbols. By a symbol we mean a portion of a drawing which is used frequently. It is therefore stored in a separate file so that a copy can be retrieved whenever needed and put into the drawing. By its nature the various entities in the symbol belong together and need to be moved together when being repositioned so that it is natural to make a symbol a group. Normally, the binding in a group is achieved in the data of the drawing by marking the data in some suitable way. For a symbol the items can be bound in a different way which has some interesting properties. The obvious way to put a symbol in a drawing is to copy all its entities into the drawing data as if the user had just created them himself at the selected position, but there is another way of doing it. Instead, a small amount of data can be put in which simply says what symbol is to be placed at the particular location in the drawing and what file contains the full details of the entities composing it. The full details are never copied into the drawing. They are just read when it is time to output the drawing to the screen or plotter. This way of doing symbols has three important properties. Firstly, it saves space in the drawing data. Secondly, any changes in the symbol are retroactive: they will appear in drawings done before the symbol was changed. Thirdly, the entities in the symbol are bound together simply by being stored outside the drawing where they cannot be touched.

Attributes

An attribute is any item of non-geometric information attached to an entity. The entity is usually a group or a symbol. The attribute can be a text string specifying, for example, a finish, a material or a part number, or it can be a number giving such information as density, cost or quantity. CAD/CAM systems can vary widely in the extent of the attributes provided from just one text string to complete pre-defined specification formats made up of text strings, numbers and multiple-choice lists. Attributes are of little use without good software for searching the drawing file and printing out tabulated lists formatted to the user's requirement.

ENTITIES AIDING DESIGN

In the remainder of this chapter we deal with some items which do not actually define any part of the design themselves but which are often stored with the design because they hold information which greatly aids the design process. Some software treats them as entities like all the others and holds them in the CAD drawing. Other software stores them separately.

Windows and screens

It is extremely useful to be able to see more than one view of the design on the screen at the same time, particularly if the resolution of the screen is not particularly good. For 2D software it is useful to see the whole design in one part of the screen and an enlarged view of the portion one is working on in another part. With 3D software, one needs to see several orientations of the geometry as well, both to aid visualisation and to access parts obscured in other views. Since many different views are possible and it may take time to set up a particular one, it is convenient to be able to store the parameters of a view once it has been set up. The parameters are often called a *window*.

Having defined several windows there may be a particular arrangement of windows on the screen which one would like to recall at a moment's notice. Some software allows the entire arrangement to be stored for recall later and it is often termed a *screen*.

Axes

Every CAD drawing has a single origin for its coordinate system. This can be placed in any particular position with reference to the actual drawing, as it is the position of the various lines relative to each other that matters. Some systems do not even make it visible but assume that the designer is going to build up his geometry on a set of construction lines. The origin is there purely because the method of storing the geometry requires an origin somewhere. Other systems assume that the designer is going to make frequent use of explicit coordinates. If this is so he will need to see where the origin is and how the axes are orientated, which calls for a visible axis symbol. Furthermore, it is often convenient to specify coordinates against some local coordinate system positioned and orientated in relation to a particular item of geometry. An example might be a lever which is at an angle to its mounting. Holes and bosses on the lever are more conveniently located relative to its centre line which is rotated with respect to the rest of the geometry. To be able to put down a second set of axes which can be turned on and used as a temporary origin whenever one is working on the item is a convenience.

Transformations

A transformation is a mathematically defined alteration to some part of the geometry. Typical transformations are linear movements, rotations and magnifications. All comprise a factor giving the magnitude of the movement or whatever. Since new geometry is often defined in terms of a transformation, the magnitude generating some key dimension, it is convenient to be able to store the transformation in order to repeat it on some related item. Transformations are described in detail in "Transformations" on page 113.

EXERCISE

What types of entity would you use to define a flat washer sufficiently well for the CAD system to calculate its volume? What additional types of entity would be needed to allow a specially written program to extract full manufacturing details about it?

Chapter 10 Geometric constructions

The geometric construction facilities need the least explanation since they parallel what is already done on paper for the most part. However, they often do the job in a somewhat different way since the software has a much more dynamic and flexible medium than paper to work in. For example, a circle through a fixed point with its centre on the cursor may expand or contract on the screen as the cursor is moved: erecting a perpendicular may be done by literally rotating a line through a right angle.

Furthermore, a degree of intelligence can be put in the software so that it infers the user's intention or makes sensible choices in ambiguous situations. For example, the software may either put a point down exactly at the location of the cursor or at the intersection of two lines if they are within a certain distance of the cursor. Intelligence has the effect of simplifying or reducing the commands the user has to give. The various programs differ in the degree to which inference is used but the development of artificial intelligence technique will increase its use. Software making extensive use of intelligence can appear to have a rather limited repertoire of commands because each command has many different results depending on the circumstances in which it is used. An example is the trim command which many programs provide. It asks the user to select a line and a second element which could be a point, line or arc. The line is then trimmed back to the second element. If it is a line intersecting the first then the trimming is done to the intersection (Figure 10.1).

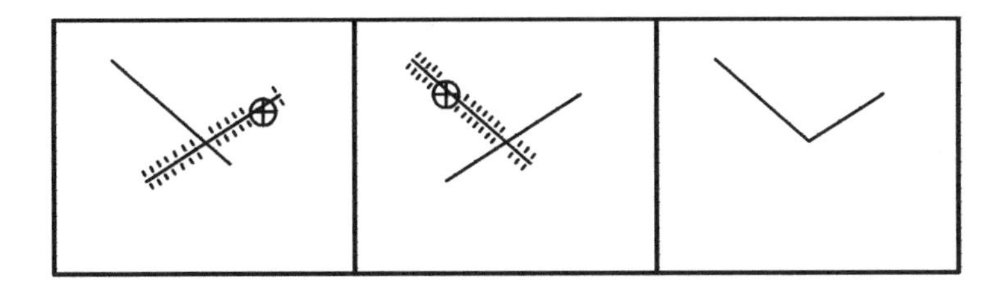

Figure 10.1 Intelligent trimming - back to intersection

If they do not intersect then it is trimmed back or extended to the point where they would intersect if extended (Figure 10.2).

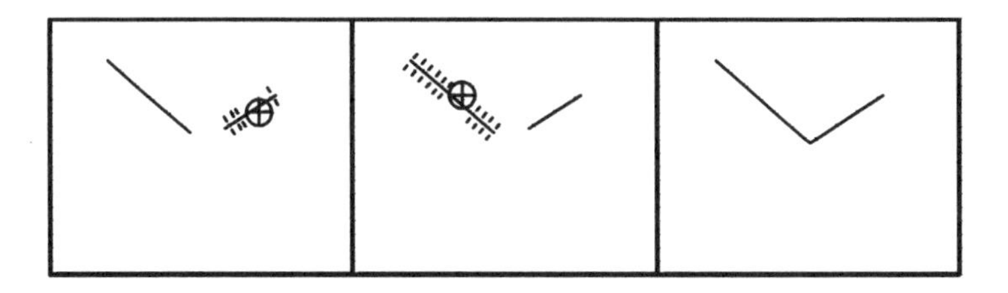

Figure 10.2 Intelligent trimming - extend to intersection

If the second element is a point not lying on the line then the line is trimmed to the foot of the perpendicular from the point (Figure 10.3).

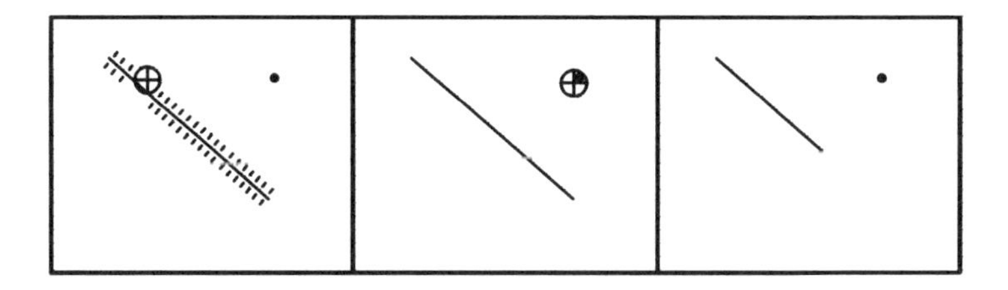

Figure 10.3 Intelligent trimming - foot of perpendicular

Within the command the software selects one of many different possible calculations depending on the elements it is given.

POINTS

All construction commands use points as inputs and many programs provide versatile, fast and efficient means of specifying points. The position of the cursor can be used in any of the following ways to define a point:

- Exact position of cursor (Figure 10.4)
- Nearest intersection of two lines (Figure 10.5)
- Nearest tangent point of a line to an arc (Figure 10.6)

- Nearest end of a line (Figure 10.7)
- Midpoint of nearest line (Figure 10.8)
- Centre of nearest arc (Figure 10.9)
- Nearest grid point (Figure 10.10)

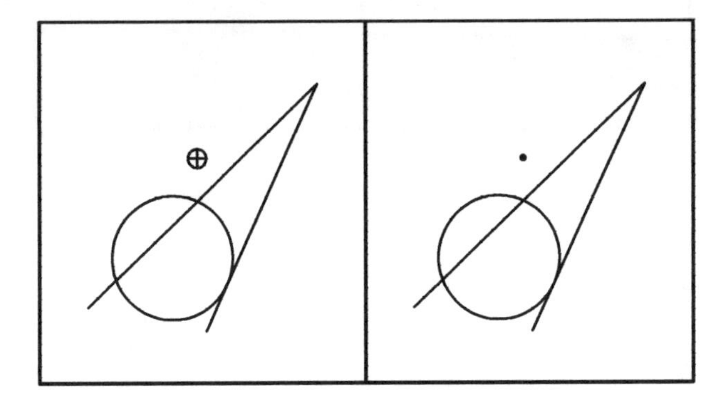

Figure 10.4 Defining a point - exact cursor position

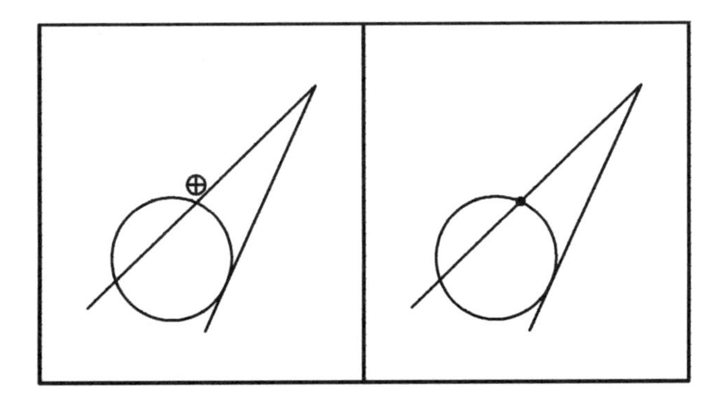

Figure 10.5 Defining a point - nearest intersection

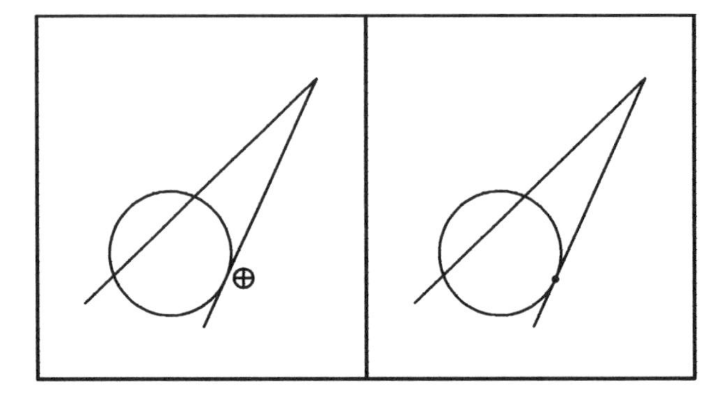

Figure 10.6 Defining a point - nearest tangent point

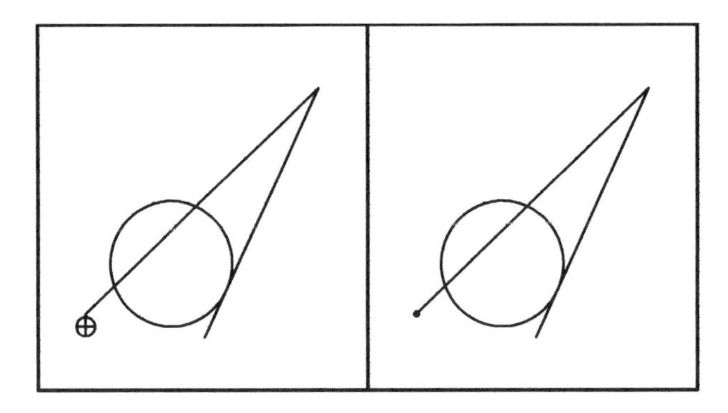

Figure 10.7 Defining a point - nearest line end

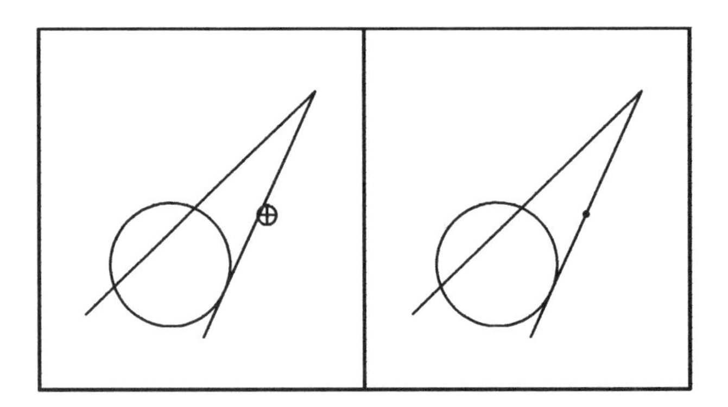

Figure 10.8 Defining a point - nearest midpoint

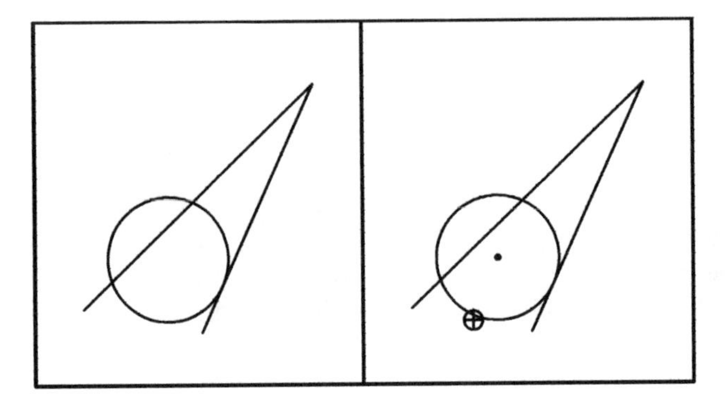

Figure 10.9 Defining a point - nearest centre

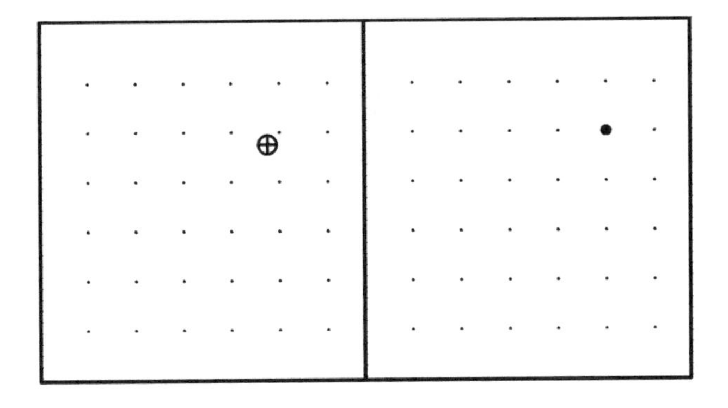

Figure 10.10 Defining a point - grid point

All except the first are often called *snaps* because the position jumps to an existing point near the cursor. Some programs require that the user specify which type of snap he requires while others select the type of snap according to what is nearest the cursor, resolving any ambiguity according to a priority rating in the list of snaps. In fact, most of the time it is the nearest intersection or the nearest end point which the user requires so that automatic selection of the snap is a very efficient technique.

Grids are a useful feature in diagrams, illustrations and schematics. When a grid is in use the drawing is covered with a rectangular array of equally spaced grid points which may or may not be visible. Any position indicated by the cursor is forced to coincide with the nearest grid point. This ensures that items are aligned and that vertical and horizontal lines are exactly so when drawn freehand. Variants on the theme are grids which have a different spacing horizontally and vertically, and grids which have points on 30° and 60° lines to aid perspective drawing.

ARCS

There are many ways of constructing arcs of circles. Those which do not use a point at the centre can be tricky. It is always useful if the centre is available as a point to use for other constructions however the circle was generated. Fillets between lines, arcs through three points on the circumference, and arcs specified by two end points and a radius are all arcs generated without using centres. The latter type has four alternative solutions, namely the centre on either side of the chord between the points and the minor or major arc through the points for each of the two positions of the centre as shown in Figure 10.11. The user has to specify which of the four he wants.

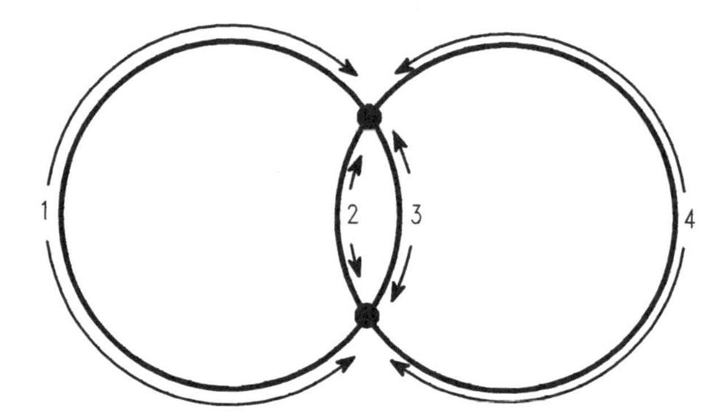

Figure 10.11 The four arcs through two points

THE OVERLAY OR BOOLEAN TRIM

A useful feature which is provided in surprisingly few programs is the overlay or Boolean trim facility. It is a powerful trim operation for cases where one closed outline overlaps another, representing one object in front of another. The user specifies which outline is "in front" and the program trims from the other outline all parts lying inside the first as shown in Figure 10.12. An extension of the facility allows the user to specify any of the Boolean operations, union, subtraction or intersection. The resulting shape is one which encloses all space which is in either, is in one but not the other, or is in both according to the operation selected.

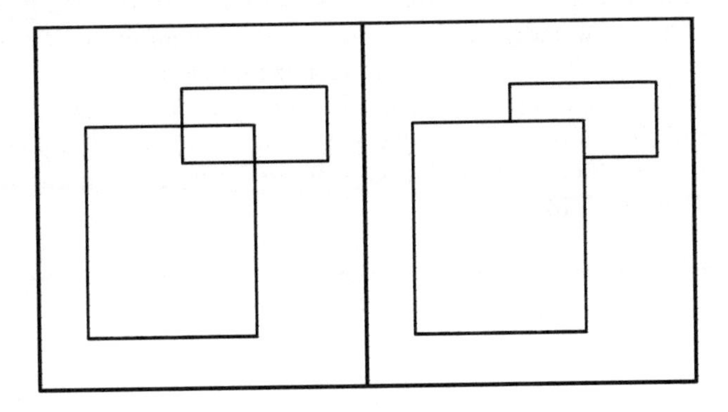

Figure 10.12 The overlay trim

OFFSET

Since engineering design is concerned with physical objects which have thickness and are often made of sheet or plate metal, an offset function, shown in Figure 10.13, is useful. In its simplest form it generates a set of lines parallel to an existing set of lines. In more intelligent implementations it puts a radius or a chamfer on the corners.

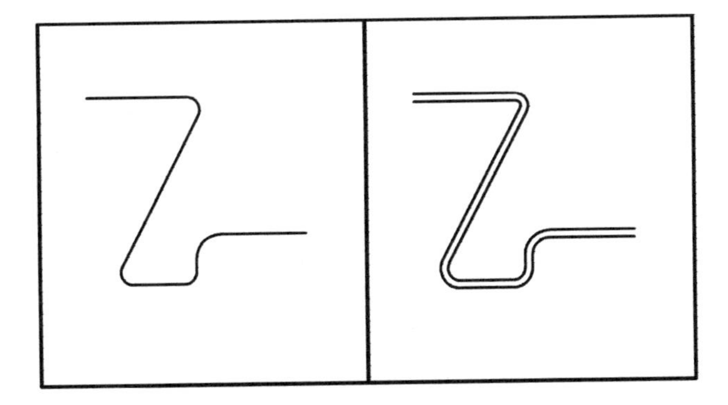

Figure 10.13 The offset line

TWO-DIMENSIONAL PROJECTION AIDS

A useful facility in 2D CAD programs is assistance in producing projected views. While not actually holding a three-dimensional model the program can obtain data defining the geometrical relationship between the planes of the various views. To do a projection the user defines two lines representing depth limits in the projected view. Wherever he identifies a point or curve in the other view, a line is generated between the two limits as shown in Figure 10.14.

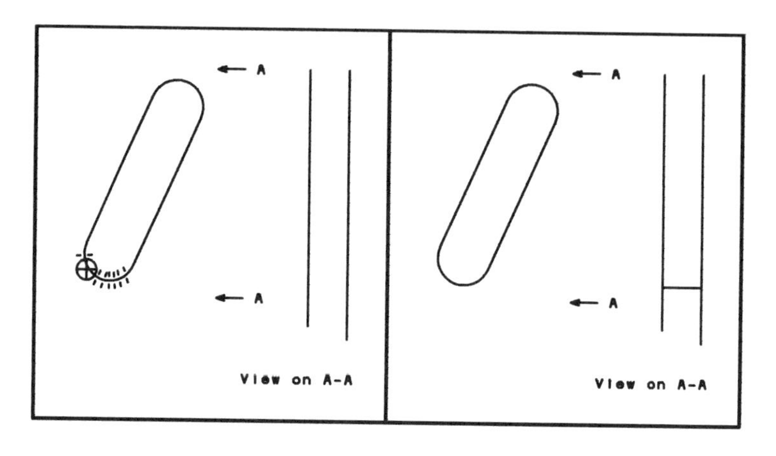

Figure 10.14 Projection in 2D

A full implementation of projections allows the user to identify by name as many views as required and retains a record of which view each geometric entity belongs to as well as the geometric relationships between the views. A simple facility may just provide the means of generating an orthogonal projection when required and keep no record of the different views.

THREE-DIMENSIONAL TECHNIQUE

Despite the additional degree of freedom the geometry still has to be generated by positions of a cursor on the two-dimensional plane of the screen but the cursor is no longer a point but an infinite line perpendicular to the screen. In selecting items in space it cannot distinguish between items situated behind each other and it cannot specify a point in space on its own. To do that the line it represents needs to intersect another entity such as a plane or a line. Geometric construction is therefore greatly assisted by being able to restrict temporarily the new work to a particular plane, and for this it is useful to keep planes as geometric entities and to have a good range of facilities for defining them.

Apart from laying geometry down on planes, one can construct three-dimensional geometry by building on the existing lines and points or by specifying coordinates. The use of temporary coordinate systems is an advantage. There is a number of constructions peculiar to three-dimensional geometry such as projecting one line on to another to generate a point (Figure 10.15), or projecting a line on to a plane or surface to generate another line (Figure 10.16). Points can be generated by intersecting lines with planes, and lines by intersecting planes or surfaces with each other.

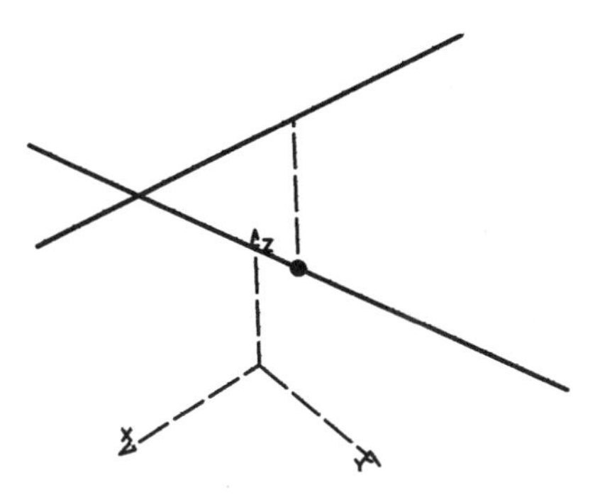

Figure 10.15 3D projection of a line on to a line

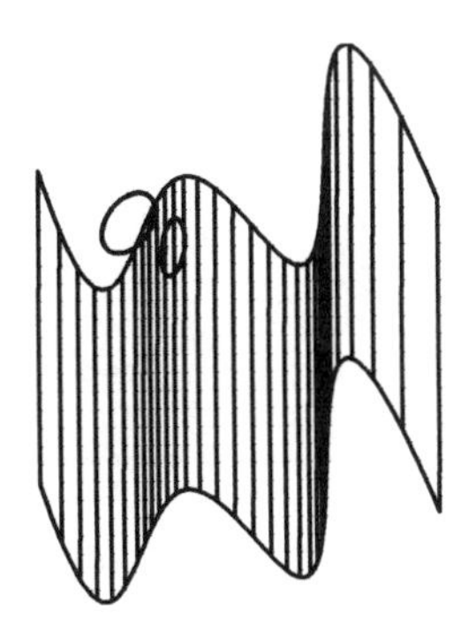

Figure 10.16 3D projection of a line on to a surface

THE BEST REPERTOIRE

The tools provided for paper drawing are limited to what is mechanically possible: compasses, parallel rulers and protractors. The tools that software can provide are limited only by the ingenuity of the software writer and the ability of the user to learn them. What is the best repertoire of operations then?

One aspect which the newcomer to CAD does not discover until he actually starts using it seriously is that it is not enough to understand what each command does. One has to learn how to put the operations together to achieve the desired result. The strategy is as important as the tactics and several strategies or methods are possible as detailed in the following sections.

The construction line method

This is nearest to the paper drawing technique. The designer draws straight lines marking out the edges of the outline. The intersections of these lines then mark the corners of the outline or other key points. CAD software supports this method by providing commands to create straight construction lines of a special type reaching to the edges of the screen at specified distances from the origin or existing elements and at specified angles. The final outline is then drawn over them using a normal line type picking up the intersections. The construction lines are then erased. For example, a circular arc would be constructed by drawing a circle and two lines cutting it as construction lines. The arc command then uses the intersections of the lines with the circle and the centre of the circle as defining points as shown in Figure 10.17.

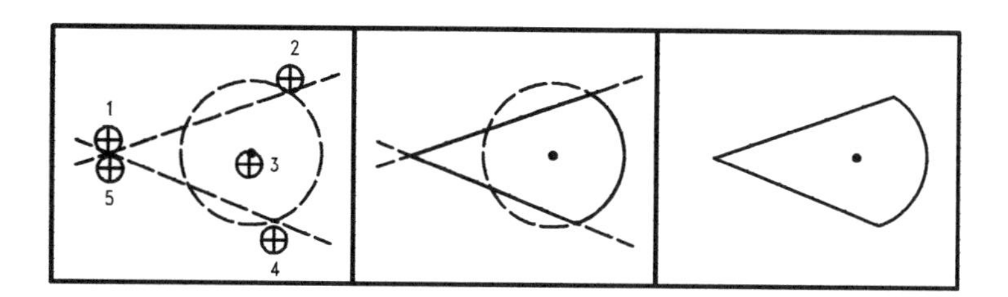

Figure 10.17 The construction line method

A variant of the method is to trace over portions of an existing shape, selecting various vertices as required to specify the positions of points. If the new geometry is on a separate layer[1] it can be easily separated from the existing geometry. Some systems allow a whole drawing to be brought in specifically for this purpose as an "overlay".

The trimming method

The user lays down long lines as in the construction line method but using the normal line type. The outline is produced by trimming these lines back to their intersections. A circular arc would be constructed by drawing a circle and two lines intersecting it. The trim command is then used on the circle to cut it back to its intersections with the lines as shown in Figure 10.18.

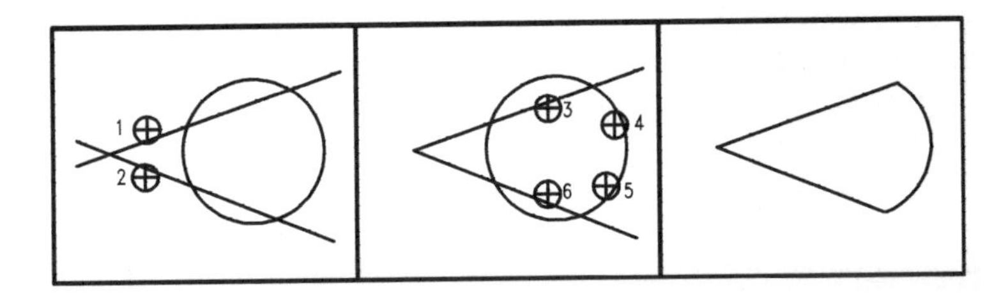

Figure 10.18 The trimming method

The navigation method

Here, each successive line segment of the outline is specified as a length and an angle from the end of the previous one. The software supports this by providing a means of specifying relative coordinates in either Polar or Cartesian form. The circular arc would be constructed in this method by feeding relative Polar coordinates into the arc command. Starting at one end of the arc the centre would be located as being a distance equal to the radius away from the end and in a particular direction. Having established the centre the other end is located as a distance equal to the radius away and in another particular direction as shown in Figure 10.19.

---

1 Layers are described in "Selection facilities" on page 111

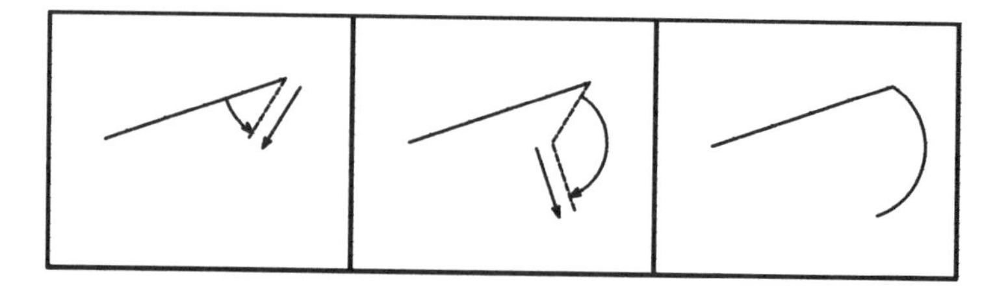

Figure 10.19 The navigation method

The parameterised method

This is only possible with CAD. The user selects a shape such as a rectangle or arc and supplies the key parameters such as length and width or radius and angles. The circular arc would be constructed in this method by the user requesting an arc and keying in the radius and the angle of each end with respect to the centre. Having created the shape the user indicates its position in the normal way.

The transformation method

This is also a pure CAD technique. The user achieves the construction by applying rotations, translations or magnifications to existing lines and points. For example, to draw a perpendicular bisector to a line he might superimpose a copy of the line on top of itself, magnify the copy with respect to one end by 0.5 and then rotate it about the other end by 90° as shown in Figure 10.20.

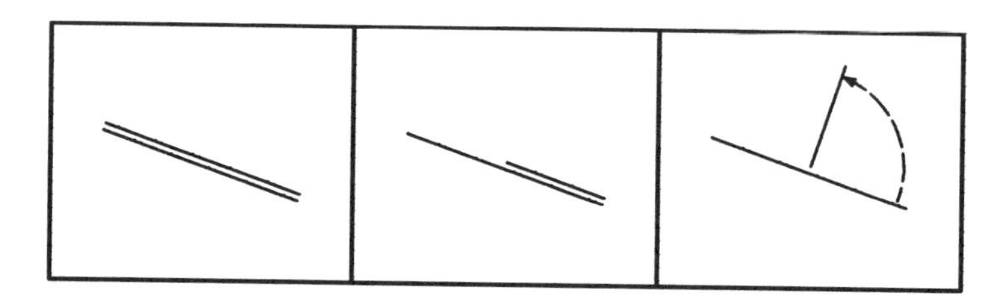

Figure 10.20 The transformation method

The subsidiary axis method

Here the user is allowed to define a temporary coordinate system displaced and rotated in relation to the normal one. For instance, dimensions of parts are often specified from some datum on the part, or one part with parallel square sides is sometimes rotated with respect to another. The use of subsidiary axes avoids much complex construction in these cases.

STRATEGY AND COMMAND REPERTOIRE

As can be seen from the above, there are many ways of achieving the same result. To return to the question about the best repertoire to use, the repertoire should be designed to make one or more of these methods efficient. If trimming is to be used then a really versatile trim command must be present. If construction lines are the main method then besides a recognisable construction line type which can be easily deleted when no longer required there must be a fast way of finding intersections between lines, and all commands requiring points as input must accept the nearest intersection to the cursor as input. The navigation method requires an intelligent keyboard input facility which can distinguish between absolute and relative Polar and Cartesian coordinates, even accepting mixtures of typed coordinates and cursor positions. The transformation method needs easily and rapidly specified transformations which will move line ends separately as well as whole lines bodily.

In looking at different CAD systems, one can usually see that the command repertoire has been designed with a particular construction strategy in mind.

SURFACE CONSTRUCTION

There is a number of well-known surfaces such as planes, cylinders, cones and spheres which are most conveniently created by simply supplying parameters (radius, length, etc) and a position and orientation. For general curved surfaces of arbitrary shape the general principal uses the fact that a surface is generated by moving a line through space. There are four things one can control in designing the surface:

- The shape of the line
- The shape of the path the line follows in space
- The way the orientation of the line varies as it moves through space
- The way the shape of the line varies as it moves through space

In the simplest type of surface generation the line is straight and follows the arc of a circle, thus generating a portion of a cylinder. If we make the ends of the line run along two curves in space then the length and orientation of the line varies as it moves. These types of surface are called ruled surfaces. Next, one can take a curved line and slide it along another curve. A particularly simple version of this is to take a closed planar curve and slide it along another curve so that a kind of pipe is produced. Alternatively, one can slide a curve along two other curves defining its ends. Finally, one can arrange by some means that the curve itself changes shape as it moves. The latter case provides the greatest generality of surface. A common technique is to place the moving curve, or generating curve, in a plane, known as the generating plane. This generating plane is then made to slide along another curve, known as the spine, so that it stays perpendicular. The spine thus controls the orientation of the generating plane as it moves. The shape of the generating curve can be controlled in a number of ways. A series of sections can be placed in space and the shape interpolated as it moves from section to section. Alternatively, a number of control curves can be placed in space and the control points of the generating curve made to follow the intersections these curves make with the generating plane as if the control points were using these curves as railway lines.

EXERCISE

Name the construction techniques you would use in the following cases:

1. A block diagram

2. Folded sheet-metal parts

3. Curved sheet-metal parts

Chapter 11 Selection facilities and transformations

SELECTION FACILITIES

The productivity benefit of CAD is obtained largely by the ability to reuse previously drawn geometry. To do this, one must be able to select the particular piece of the drawing one wishes to use. Facilities for making the selection are therefore important. The value of a good range of selection methods will be realised once you need to pick up an outline buried in a drawing densely filled with lines and texts all over the place! It is surprising how poor a range of selection techniques is sometimes provided. Yet there are potentially many methods as the following list shows:

Boxes A rectangle is drawn round the elements to be selected. As a better alternative, a polygon may be used. An additional parameter may also be accepted to indicate whether the selected elements are to be those which are wholly in, partially in or wholly out of the box. This technique is the one most commonly provided. Sometimes the user is allowed to have more than one box.

Types The user specifies that all lines or all circles or all texts etc shall be selected. This is really of not much use unless it can be used as a qualifier to some other selection in force at the same time.

Names Some software allows the user to give names to individual entities. He then performs the selection by specifying one or more names or a wild-card name but a powerful alternative is the ability to collect an arbitrary set of entities together under a single name supplied by the user or generated by the software. The facility is even more useful if any entity in the group may itself be such a group so that a "family tree" can be organised in the design. Specification of the name of the group then selects all the entities in the group including those in groups contained in the group.

Markers

A very easily implemented facility is to give each entity a binary on/off marker. The user sets it on for the entities he wishes by selecting them with the cursor. The selection is then applied to all entities with the marker on. The markers would normally be reset after the selection to avoid confusion. An elaborate and powerful extension of this technique is to use the other selection methods to set the markers and accumulate a set of marked items by a series of different selections. The method becomes even more sophisticated if selecting the item turns the marker on or off depending on whether it is already on or not. It is then possible to build up a complex selection condition according to the rules of Boolean algebra - for example "all lines inside the box but not on layer 5".

Layers

It is common to provide each entity with an integer (sometimes a name) called the *layer*. Layers model a stack of transparent sheets laid on top of each other in registration, each holding a different aspect of the design. One of the layers may be chosen as the current layer on which all new geometry is put. One or more of the other layers can be made accessible for selections, entities on the remainder being unselectable and invisible.

Elaborations on this facility sometimes provided are separate controls over whether a layer is visible or not, whether entities on it can be selected or not, and whether entities can be altered or not.

THE USES OF SELECTION

As stated at the beginning of the chapter the main use of selection will be to take a copy for further use. It can also be used to delete parts of a drawing selectively. Again, since some items have parameters such as the height of text or the thickness of lines, it may be useful to be able to set a particular parameter on a whole collection at once. Another very useful operation is transferring collections of items on to a particular layer. Last and by no means least is the powerful group of geometric editing functions known as transformations. These will be described in the remainder of the chapter.

TRANSFORMATIONS

In a transformation, a selected part of the geometry is subjected to a precise mathematical alteration in order to reposition it or make it a different shape or size.

Transformations are particularly useful where you first make a copy of the selected portion and then transform the copy to produce an addition to the drawing. In this way, new geometry is derived from existing work without redrawing. You never draw the same shape twice. It is this ability to use a previous piece of drawing to form a new piece, so that one never draws the same thing twice, which makes transformation facilities one of the principal ways of gaining productivity, second only to parameterised design in value.

There are three parts to performing a transformation: selecting the items, optionally copying the items and then actually altering the items.

TYPES OF TRANSFORMATION

CAD software varies in the range of transformations offered. Translation, rotation, magnification and mirroring about a horizontal or vertical line are usual. Others sometimes provided are shearing and mirroring about other axes or about a point.

Translation

In translation the selected entities are simply moved from one point to another without rotation. Its uses are obvious. In modifications it allows one part to be moved over to make room for another part, or a part to be lengthened or shortened. In new designs it allows a pattern of holes, say, to be repeated or even just a distance between two ends, for example, to be used several times over on different parts.

In performing a translation, a vector has to be defined to specify the direction and the amount of movement. This is done by one of the following:

- Selecting two points
- Selecting a line
- Keying in an angle and a distance

It should be noted that the lines or points can be anywhere in the drawing and need not be associated with the geometry to be moved, in which case the operation can be regarded as copying a vector (a length plus an angle) from one part of the drawing to another (see Figure 11.1).

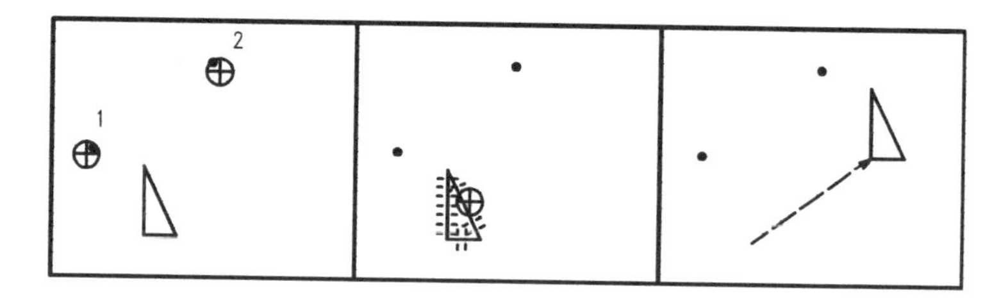

Figure 11.1 Translation

Rotation

Rotation is clearly of value in designing all the many circularly symmetric components used in engineering. Besides its obvious use in changing the orientation of something it can be used to input the angle of one item relative to another by drawing one on top of the other and then rotating one of them. A typical example might be the setting of holes at equal angles round a pitch circle. As mentioned in "Geometric constructions" on page 97, one way of erecting a perpendicular is first to superimpose the perpendicular on top of the line and then rotate it through 90°.

Rotations require in 2D:

- A pivot point
- An angle

Rotations require in 3D:

- A pivot point
- Two angles

or:

- A line
- An angle

These are illustrated in Figure 11.2.

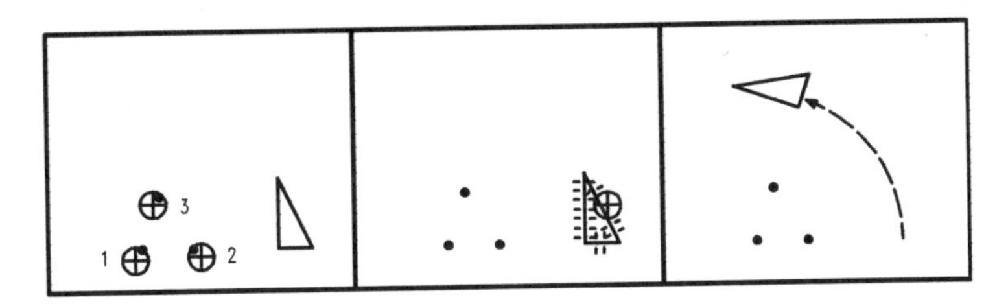

Figure 11.2 Rotation

The angles can be defined by:

- A numerical value
- Two other points subtending the required angle at the pivot
- Three other points

There are thus many possible ways of defining a rotation.

Mirroring and other symmetry transformations

CAD sales demonstrations often include a mirroring operation, presumably because it is spectacular and is not at all easy to do by other means. There are so many symmetrical parts in engineering that it has a clear value for design and draughting.

To do the operation, one needs only to specify a line to serve as an axis about which the geometry is to be mirrored. This can be done by specifying a point and then saying whether mirroring about a horizontal line or a vertical line through the point is required as shown in Figure 11.3. More elaborate software offers the choice of a line at any angle as shown in Figure 11.4.

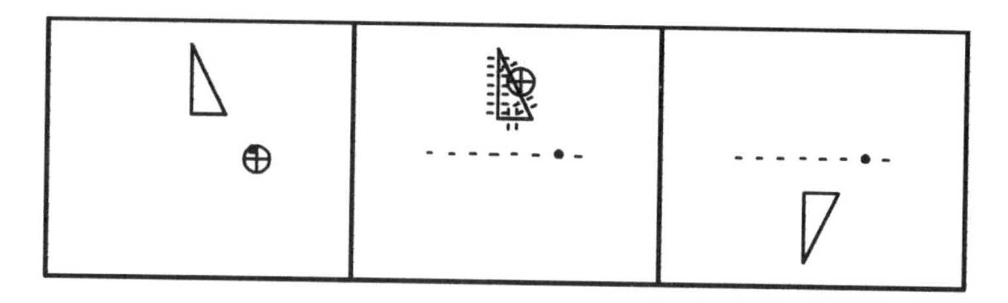

Figure 11.3 Mirroring about the horizontal

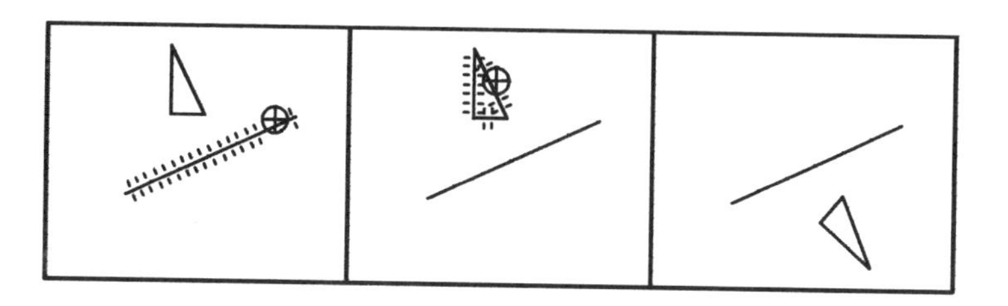

Figure 11.4 Mirroring about any line

If this function is not available, mirroring about any line can be achieved by marking a point on the line, mirroring about a horizontal line through the point and then rotating about the point by twice the angle the line makes with the horizontal as shown in Figure 11.5.

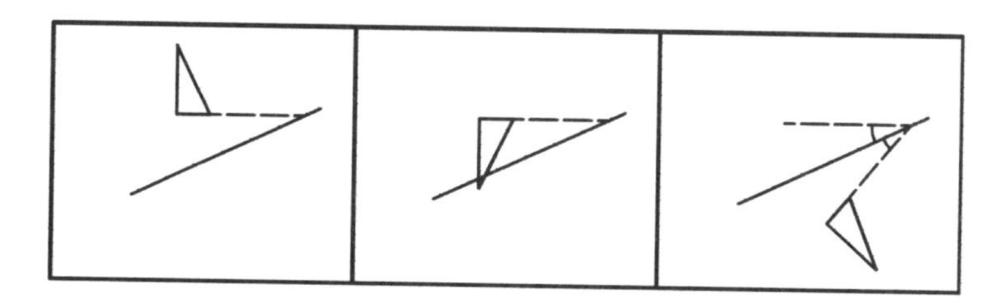

Figure 11.5 Mirroring using mirroring about the horizontal

Another elaboration is the oblique symmetry. In simple mirroring, each point is moved perpendicular to the mirroring axis to a position the same distance on the other side of the axis. In oblique symmetry, each point no longer moves in a direction perpendicular to the axis but in a direction specified by the user. The effect is to shear as well as mirror as shown in Figure 11.6.

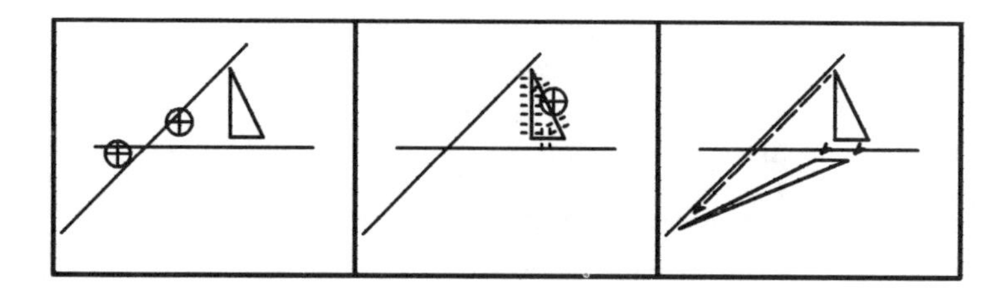

Figure 11.6 Oblique symmetry

In 3D systems, normal mirroring about a line is equivalent to rotating 180° around the line whereas mirroring about a plane has the same reversing effect as mirroring about a line in 2D. Mirroring about a point is equivalent to a 180° rotation about the point.

Magnification

This is normally taken to include reduction since the operation takes a factor which can be less than one as well as greater than one.

Magnification has an obvious use in scaling parts up or down when designing larger or smaller variants. A less obvious use is to subdivide a line by magnifying a copy of the line by the reciprocal of the number of parts.

To specify a magnification you need to:

- Select a point about which the magnification is to take place
- Define the magnification factor

The factor can be specified by:

- A number
- Selecting two points such that the ratio of their distances from the pivot point is the required factor

The use of two points is equivalent to asking that the magnification be such as to move the position of the first point to that of the second. Some software interprets this literally so that if the points are not co-linear with the pivot a rotation occurs as well.

Differential magnification

This is a variation of magnification sometimes provided which is particularly powerful by virtue of its ability to alter the shape of things. What is offered is a different magnification factor in the horizontal direction to that in the vertical direction so that squares can be turned into rectangles etc. As before, a pivot point has to be defined but two factors are required instead of the single one (see Figure 11.7).

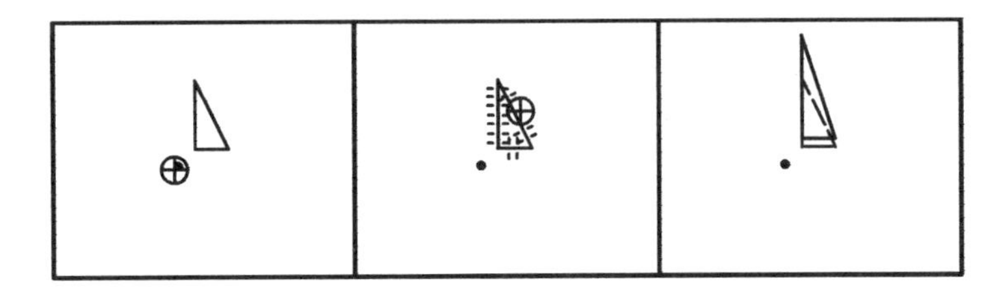

Figure 11.7 Differential magnification

A particularly useful case of differential magnification is in stretching a rectangular plate, for example, by a certain factor in length while leaving the height unchanged. In this case the vertical factor is set to unity.

Shearing

Like differential magnification this changes the shape of an outline. It is not often provided. A shear is defined by a datum, an angle and an axis. All points are moved in the direction of the axis. Figure 11.8 shows a rectangle being sheared.

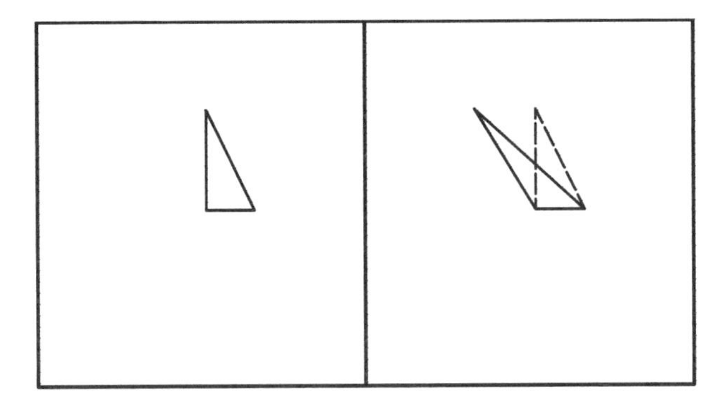

Figure 11.8 Shearing

Multiple transformations

The user is sometimes allowed to specify several transformations in a single command. A simple example has already been mentioned where specifying a pivot point and two other points determines both rotation and magnification, the geometry being transformed so as to make the two points coincide while keeping the pivot point fixed. The concept can be extended. If two points corresponding to the old version are specified followed by two points corresponding to the new version and the geometry is transformed so as to fit the first two on to the second two, a translation, magnification and rotation are all specified in one command. The method can be extended yet further with three points in the old geometry being fitted to three arbitrary new points. The result will be translation, magnification, rotation and shearing with mirroring if the third point is mirrored with respect to the other two. The technique is powerful and particularly convenient if the user is allowed to cut short his specification of the three new points to either one or two if he wants just translation or rotation. Figure 11.9 shows a three-point transformation.

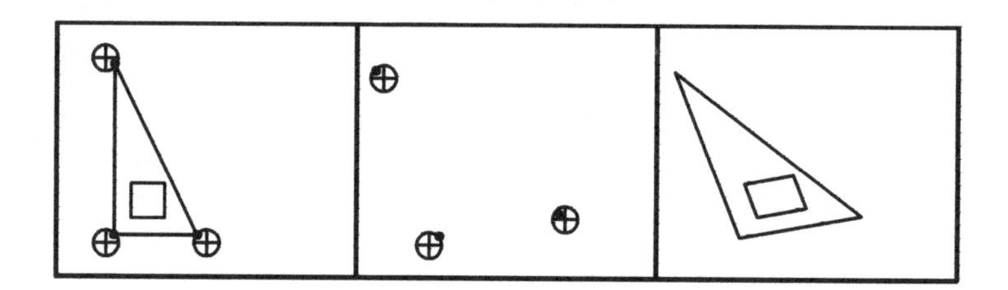

Figure 11.9 Three-point transformation

Another way of specifying multiple transformations is to specify another set of axes. This describes a translation and a rotation.

COPYING

As was stated earlier, transformations are particularly useful if they are performed on a copy of the selected geometry as this allows the creation of new geometry from old without redrawing. Copying can be provided as an option on the transformation, the user choosing whether the transformed geometry is to replace the selected geometry or to be an addition to the drawing.

Alternatively, copying is provided as a separate facility in which the selected geometry is stored away in a buffer or a disk file. In this form it provides a symbol library facility which allows the user to keep a library of frequently used items for inserting into drawings. The user is then allowed to apply the transformation to the geometry selected from the library just before it is actually inserted into the drawing.

Whenever geometry is stored away for future use, a point, usually known as a datum, has to be defined. Geometry retrieved from store is positioned in the drawing by selecting a point. The retrieved item is then located with its datum on the selected point. Transformations applied before it is inserted take place relative to the datum. With three-point transformations, three datums are defined before storing. When the item is retrieved, three location points may be selected, thus defining position and transformation.

TRANSFORMATIONS ON LINE ENDS

Some programs treat lines as completely self-contained entities. The transformation affects the line as a whole, moving it bodily. A selection box cutting the line in two will either select the whole line or not depending on the option in force. Other programs treat a line as an elastic band joining two points, each of which may be separately selectable. A box cutting the line in two will then select just one end of the line and move the point about according to the transformation specified. The effect is to make a wider range of distortions available. The facility is essential in circuit schematics where the user often needs to move components around without breaking the lines connecting them.

EFFECT OF TRANSFORMATIONS ON TEXTS

If you mirror a section of a drawing containing some text the text would become unreadable if it was mirrored as well. Text is usually represented by a single datum in the geometry around which a subroutine draws the lines which form the characters according to various parameters such as height etc. All that is required when mirroring a part of the drawing containing text is that the datum stays in the same place relative to the rest of the part being mirrored. So the transformation must be applied to the datum only. But what if the part is being magnified? Should the text height be magnified as well? Sometimes this will be wanted and sometimes not. The same question arises with rotation.

We need to be able to supply some parameter which says whether the transformation is to be applied totally or just to the datum of texts. Another alternative is simply to require that the text should remain legible: meaning that the text is never written upside down, is always readable from the right-hand side of the drawing and is a legible size. This is illustrated in Figure 11.10.

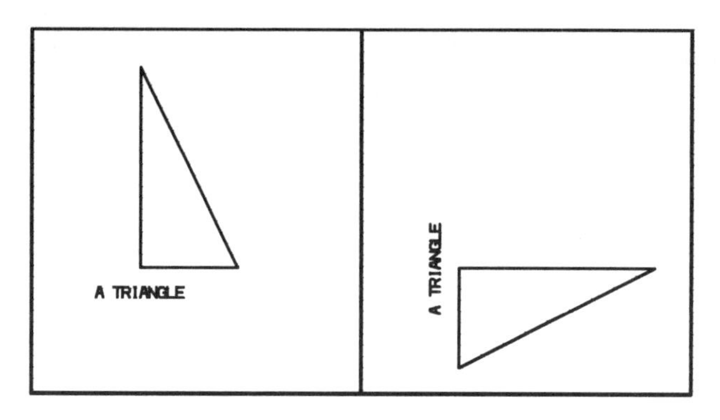

Figure 11.10 Legible rotation of text

Similar considerations may apply in a lesser degree to symbols in which the detailed geometry is described outside of the drawing, the only data in the drawing itself being a datum point, a rotation angle, a magnification factor and a descriptor pointing to the place where the geometry is held.

EXERCISE

Give the transformations you could use and how you would use them in drawing the following:

1. A turned component

2. A spoked wheel

3. Lengthening a lever with a hole at either end

4. A view of a portion of a part shown at a different scale

Chapter 12 Dimensioning, annotation and graphic effects

We have drawn attention occasionally in previous chapters to the great value of dimensioning operations in CAD. First and foremost they reduce the risk of human error by putting on to the drawing the actual calculated length or angle and not what the designer thought he meant. Secondly, they automate quite a bit of actual drawing. The designer indicates the length to be dimensioned and the position for the written value. The software generates the written value and tolerance and all the related lines and arrows. It is a pity that (certainly at the time of writing) dimensioning facilities are the ones most often skimped in a CAD program when it is first launched. I have even seen facilities which would barely pass muster in one of the most expensive packages on the market! The trouble is that dimensioning is not glamorous, has to cover a wide range of different styles, and requires a lot of tedious detail to implement and a lot of knowledge of drawing practice. None of this is attractive to software writers trying to launch a new program in a competitive market where the latest fashionable feature and buzz-word counts. Nevertheless, the range of styles should meet the requirements of the appropriate national or international standard.

The degree to which the dimensioning is automatic has varied widely from software which simply provides the value as a text and leaves the user to position it with lines and arrows at one extreme to software which puts a dimension on everything in sight with a single command! The usual provision is a command in which the user identifies the distance to be dimensioned and the position of the value while the software does the rest. In some systems the software will work out a position for the value or space out the values evenly where they have to be stacked on top of each other. If the value is positioned automatically then it is usual for the software to put it outside the witness lines if it will not fit between them.

The nature of automatic dimensioning requires good facilities for customising it to the requirements of a particular company. A list of desirable customising parameters can be found in the checklist in Appendix A.

Current software does not calculate the tolerance except to subtract or add a specified value in producing limits. There is a real need for a CAD system which analyses tolerances. In the mean time the only dimensions CAD will handle are the centre values. The tolerances are purely values specified by the designer according to his judgement and experience. Some software will position the two values neatly one above the other.

ASSOCIATIVE DIMENSIONING

Once you have dimensioned up, what happens if you go back and alter a length? The dimension is incorrect until you remember to redo it. Associative dimensioning is designed to overcome the problem by redrawing the dimension with the new value as soon as the alteration has been done. In order to implement it the drawing data has to hold a link (association) between the geometry defining the distance and the dimension. The dimension cannot be just a collection of lines and texts like any other lines and texts in the drawing. A problem can arise here in that it is not sufficient to link the dimension to a line since many dimensions record the distance between the ends of two separate lines. Also, where the dimensioning direction is parallel to a specified line, there has to be a link to the line as well.

ANNOTATION

The checklist in Appendix A gives the controls made available in many programs for determining how text is to be generated by the software. Control of height is essential and in this connection it is worth mentioning that plotters can produce legible text down to 1 mm high - considerably smaller than hand-printed notes and even text 0.1 mm high can be made legible on the screen by zooming in. It is very useful to be able to line up text by its left-hand end and occasionally by its centre. Being able to line it up on a point just below is useful for locating it on a line. Some programs provide extensive facilities for varying the lettering style by applying shear to the characters or altering the aspect ratio (length to height), or even allowing the user to design his own style. These are only of value if the user has a very definite reason for needing a wide range of styles. Font design is very laborious and should an engineering CAD system be used for graphical design anyway?

There are five kinds of annotation used in drawings, each of which asks for software features peculiar to itself:

1. Text with a leader line pointing to something

2. Text on its own, often multi-line

3. Geometric tolerances

4. Parts information

5. Information describing the drawing itself

Text with a leader line should preferably be associated with the element it points to so that it follows the element when moved. It is also useful if the software simply asks the user to select the item and positions the line and arrow automatically. With plain text (2) it is useful to have the software space the lines out automatically and to fit it into a box when required. Another useful feature is automatic generation of sequences of numbers since notes are often numbered. Geometric tolerances (3) are basically sequences of symbols with a leader line. If there is no special routine for them they may be produced using the symbol library or, more conveniently, by text employing special characters.

The particular characteristic of parts information (4) is that it must be possible for a program to search the drawing and extract the parts information separately for formatting into a printed parts list or for passing to other software such as a database. At the same time the designer must be able to input and position it as text. Furthermore, while it will consist of many separate items of text, certain texts will belong together as a set of attributes of one component and the program extracting the text must be able to recognise unambiguously which ones belong together and which text in each group specifies which attribute of the component - that is, which texts make up the line and which texts go in which column of the parts listing. Clearly, the texts have to be tagged somehow in the drawing data so as to make them recognisable when being extracted. Some software keeps the attributes out of the visible drawing. The disadvantage is that they cannot be plotted out as part of the drawing and the designer can only see them by invoking a special function. A good solution is to allow the user to attach an identifier to a text which says what attribute it represents. The texts describing one component can be linked together using the normal grouping method used for other entities.

Information describing the drawing itself such as drawing reference, date drawn, who by etc also needs to be readable by separate programs if a proper computerised drawing management system is to be used. It is also essential at the same time that it be visible in the drawing when plotted out.

GRAPHIC EFFECTS

Hatching and pattern filling

There is something satisfying in seeing the wave of parallel lines wash across an outline in an automatic hatching operation! Some programs are surprisingly awkward with hatching. The difficulty is in identifying the closed outline. Sometimes there are problems in identifying closed outlines inside the main outline representing voids in it and there can be problems in deleting the hatch when required. The hatch needs to be data associated with the outline instead of actual lines, as these fill up the CAD drawing with useless data.

Useful variants on hatching are filling an outline with a repeat pattern or a solid colour. In the case of a repeat pattern it may be possible for the user to define the pattern. Certain repeat patterns, such as those indicating concrete or stone, are needed in engineering drawing.

Patterned line styles

Besides the usual dotted, dashed and dot-dashed lines, there is sometimes a need for a line which has a more complicated pattern in it, such as weld lines, so many programs provide some of the more frequently used types. Another feature which may be of use occasionally is the means of defining your own pattern of dots and dashes in the line.

EXERCISE

Give the ways in which dimensioning, annotation and hatching have to be linked in the CAD data to geometric entities.

Chapter 13 Parameterised drawing and customising

A new CAD system is like a new employee on the staff. He has been brought in to do useful work for the company but however well educated or experienced he is, his usefulness only develops as he adapts to the particular ways of the company. Similarly, a CAD system needs to be adapted to the particular work it is required to do. The extent of adaptation varies from place to place, reaching its furthest in parameterised drawing where the CAD system is programmed to turn out complete drawings of the company's products. The lowest level of adaptation occurs with the customisation which is required when the system is installed.

PARAMETERISED DRAWING

Parameterised drawing is by far the biggest source of productivity in CAD. Where a range of components share a common basic shape and differ only in one or two key dimensions according to particular customer requirements it is possible to program software so that the designer supplies just the key dimensions and the software produces a whole new set of drawings. Various case studies of parameterised drawing are described in the technical press from time to time. Two good examples can be found in References (2) and (10). Reference (2) describes the effective application of macros in the CAD system while Reference (10) describes a fully implemented design system where analytical design calculations for aerodynamic and stress properties predominate.

As with all automation there has to be a net saving after taking into account the initial cost of programming the software together with the saving accumulated over all the times it will be used. The initial cost of setting up a parameterised drawing cannot be overlooked. Some kind of programming has to be done. Different CAD systems offer different ways of doing the programming, each usually claiming to be easier than the next one. But the cost is affected not so much by the power of the programming language as by the analysis and planning that must precede the writing. The design to be automated must be well understood, so well in fact that within the range of all the possible cases that may be generated any exceptional or erroneous case can be recognised and handled safely by the program. Such cases might be geometry which changes shape radically if one length becomes longer than another, topology which becomes unmanufacturable, or dimensions which become unsafe. The specification for the program should decide how to recognise them and whether to take a decision itself or ask the user for a decision to resolve the problem, or simply announce the problem and stop. A significant part of the cost will therefore be in analysis and specification. If this is not done at the beginning then it will be forced on the users during a "debugging" phase when problems appear in use and have to be corrected by changes to the program. The cost of software development cannot be avoided.

Before looking at the various ways of automating drawing it is worth while recalling the principal features of programming languages in general for those not familiar with them. A program consists of a sequence of statements. The order is important since the computer reads and obeys them (executes them) one after the other in the order they occur. Since each statement is a command to do something it will have an English word which is a verb specifying the action such as "PRINT" or "INPUT" or "STOP". In most cases the action has to be done on something so besides the verb, one or more objects are needed, or, in mathematical terminology, arguments or parameters. The arguments will be either constants (i.e. explicit numbers) or variables (i.e names representing a value which depends on data that has been supplied when the program runs). Thus, in the statement "PRINT 3.4, X1" the items to be printed are the number 3.4 and whatever value has been assigned to the name "X1". The language will also have rules for separating the verb from its parameters (a space in this example) and for separating one parameter from another (a comma here). It is usual to be able to write an algebraic expression combining variables and constants instead of a single variable if desired. Besides the verbs which do things to numbers there are verbs for altering the sequence in which the statements are executed.

Variables need not just represent numbers. They can be used to represent sequences of text characters (text strings) or lists (vectors) or tables (arrays) of numbers. This is particularly important in graphical work where, for instance, a point requires a list of two coordinates to specify it. Naming has even further scope in CAD if drawing entities can be named. The program can name items as they are created and then perform operations on them without having to use graphical selection methods like drawing boxes round them. Unfortunately, the ability to name entities is not as common as it might be in CAD systems. Naming text strings can be useful in an interactive dialogue allowing the user to give a name to frequently used constants. Seeing that commands are text strings, a command with a particular set of parameters can be given a name as well so that frequently used commands can be called up by single names. After these general remarks on software, we will turn to a review of the three different tools which can be made available for producing parameterised drawings.

The established scientific programming language

The first is the well established method of FORTRAN or a similar scientific language. It suffers from the disadvantage that scientific languages only have facilities for manipulating numbers. They have nothing for drawing and manipulating lines etc so these functions have to be obtained from a special subroutine library provided by the writers of the CAD program. In most cases the subroutines do not operate the graphical display when they make the alterations to the drawing. The designer runs the program on an alphanumeric display, engaging in dialogue with it as necessary. Afterwards he examines the result using the CAD system in the usual way. Development of the program is slowed down by the result not being immediately visible and by the function of the subroutines not being always easy to understand. By the way, the CAD software writers cannot change the functions of the subroutines at all once the library has been issued since they will be incorporated into the user's own software. The method has the supreme advantage of being able to draw on the powerful calculation facilities available in FORTRAN and its relative transferability between one computer and another. The program created runs faster than those created by the other methods.

The special alphanumeric graphical language

The next method allows programs to be developed much faster than with FORTRAN. A programming language specific to the CAD system is devised by the CAD software designers. The users write programs in the language and while they are using the CAD system, tell it to take commands from one of the programs instead. The programs can therefore be used as additional interactive commands tailored to the particular requirements of the users. The method is especially efficient if the programming language is like the normal interactive commands, so that each command verb in the language has an exact counterpart in an interactive command and the parameters needed by the command are the same as the data supplied in the interactive operation. When writing a program the user can then try out the sequence of commands first interactively. In many systems the interactive commands and the programming language are almost identical because the control dialogue actually uses the programming language. The software is designed so that drawing can be done if desired by typing commands on the keyboard. The menu tablet is set up by assigning pre-programmed commands to boxes on the menu. The user can therefore either hit boxes on the menu for the most frequently used commands, use the keyboard to type in the occasional special command or type whole sequences of commands into a file for recalling later. He may even be able to set up a "recording" mode in which a sequence of commands from the menu tablet and keyboard is put into a file while drawing is being done for use again. A related technique is to write a program in FORTRAN, BASIC or any other language of choice which writes out a file of these commands.

Parameterised geometry

The third method of parameterised drawing requires no programming at all in the conventional sense of making up a sequence of commands. It works on the principle that, since the dimensions specify the design, altering a dimension should be sufficient to alter the design. It is a kind of dimensioning operation driven backwards in which the dimension determines the geometric model instead of the geometric model determining the dimension. Furthermore, if we allow an algebraic expression to be used in the place of the number in a dimension the value of a single variable can determine a whole set of dimensions, each one in a different way according to the expression. In order to work properly the software has to keep lines joined together so that the drawing works like a mechanical linkage with lines pivoting about each other. In fact, a parameterised drawing facility of this kind can be used to simulate and solve linkage designs. For the facility to be of practical use the user must be able to choose which areas and elements will be affected.

KNOWLEDGE-BASED DESIGNS

Developments in artificial intelligence have made it possible to extend the intelligence, or decision-making, in parameterised design to the point where the program very effectively encapsulates the experience of an organisation in doing its designs. Such software is not programmed in a traditional language but uses advanced human interface techniques to allow the experience to be expressed by non-programmers directly to the software. This does not make the process of capturing the experience a trivial task but does make it a cost-effective proposition in the right circumstances which are characterised by the following:

- The design rules are actually known by some people somewhere.

- The design is highly complex such that modifying it is laborious.
- The design is repeated often for many variants. (As with all effective parameterised design.)
- The design is on a critical path for the overall project.
- The knowledge of specialists has to be made available to a wide range of people.
- The knowledge is in danger of being lost as it resides in the memory of someone about to retire.

See Reference (38) for a report of the practical application of such software. Knowledge-based design systems, being only applicable in particular circumstances, are likely to be offered as options to CAD systems. Tight coupling with the CAD system is necessary because a complete design or CAD model is generated.

CUSTOMISATION

Customisation is adjusting the software to make it fit the particular requirements of the user. There are various places where this is needed:

1. Dimensioning style
2. Text styles and sizes
3. Line widths and styles
4. Drawing frame
5. Hatching patterns
6. Symbols
7. Tablet menu layouts

The first three items in the list are usually done by supplying parameters to the software and are therefore easy to do. The remaining items can involve substantial amounts of work. A large symbol library may have to be drawn and all the drawing frames will have to be done. Tablet menu layouts are particularly difficult to get right. One does not know what the best layout is until the system has been in use for some time, by which time users have become used to them anyway!

EXERCISE

What factors, both managerial and technical, would you need to examine before setting up a parameterised design?

Chapter 14 Screen handling and output facilities

In this chapter we will consider a miscellaneous number of facilities concerned with presenting the drawing on the screen and on the plotter. We will take the graphics screen handling first.

SCREEN HANDLING

Panning and zooming

The two most frequently used operations on the screen are panning and zooming. These are required because one usually works on only a portion of the design at a time. The screen is used to present just that part enclosed by an arbitrary rectangular frame superimposed on the drawing. Panning is positioning the frame on the drawing. On simpler types of display it is done by a command which redraws the screen so as to make the cursor position at the time of issuing the command the new centre of the screen. A more convenient way which avoids a special command is to slide the drawing across when the cursor is against one edge of the screen as if the cursor was pushing the frame across the drawing. Considerable convenience is provided by displays with local processors where knobs or joysticks instantaneously move the drawing about on the screen or allow the tablet puck to "grab" the drawing and slide it about under the frame.

Zooming is changing the size of the frame, thus making the drawing smaller or larger on the screen. It does not change the dimensions of the drawing in any way, being equivalent to standing closer to or further from the drawing. The software maintains a factor, the display scale, which is the ratio between distances in the model and their size on the screen. The user will be allowed to enter a new value or, more usually, a factor (greater or less than one) to multiply the scale by. When zooming in or enlarging the display it is very convenient to do it by marking the approximate corners of the new frame using the cursor. The facility is very impressive when first encountered as the sheer range possible is far beyond its optical counterpart, the zoom lens. What appears to be a small dot on the screen can be magnified up until it turns into a message or even a complete drawing. Two lines which appear to meet can be magnified until the clearance between them is visible. The range is possible because lines have no thickness in the model and the coordinates are held to a far higher precision than is used when they are converted to lines on the screen. The coordinate system of the screen will consist of two integers lying between zero and perhaps 2000, the latter figure being the number of pixels along one side of the screen. The coordinate system in the model may consist of two floating-point numbers able to take on any value from 10,000,000,000,000 to 0.000000000000001. In converting to the screen coordinate, each real coordinate is multiplied by whatever is the current display scale. The fractional part will then be discarded to convert the result to an integer. The consequence is that all detail below a certain level is lost.

In 3D software, another important facility is rotating the model in space about some chosen axis. The whole model with its coordinate system rotates relative to the screen or, to be more precise, the plane of the screen is rotated in space around some chosen axis in the model. This is not the same as rotating an isolated piece of geometry in a transformation. The facility is essential for accessing the model when selecting points etc in order to move one point out from behind another. It is also an aid to visualisation. Furthermore, if a local processor has been provided which will link the rotation to the rotation of a knob then it is possible to get an extremely clear impression of the spatial relationships in the model by turning it about back and forth with the knob.

Projections

Representing a three-dimensional object on a two-dimensional display screen requires a particular kind of mathematical operation (projection) to turn the three coordinates of each point into the two coordinates of the screen. Two kinds of projection are used: orthographic, or parallel, and perspective, or conic. Perspective projection is what we are all used to, being the operation performed by a camera lens or the lens in the eye. The mathematical operation can be seen in Figure 14.1. As the eye looks at the screen, any point on the screen could also be a point in an imaginary three-dimensional space behind the screen lying anywhere along an infinite line joining the eye to the point on the screen.

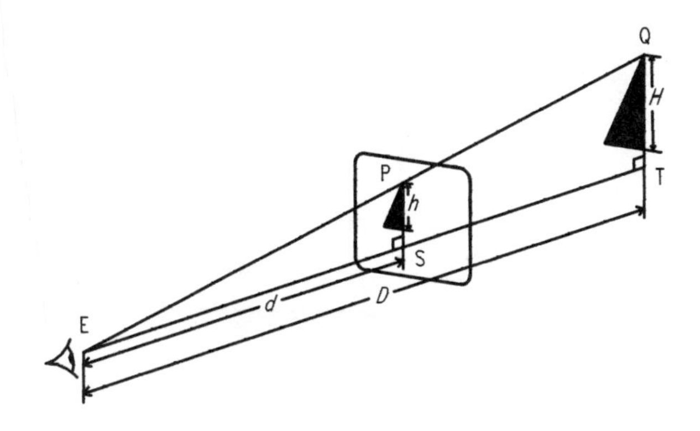

Figure 14.1 Perspective projection

To generate the projection we imagine the object to be located in the imaginary space behind the screen and draw lines from its vertices to the eye as shown in the figure. The projected points are where these lines intersect the plane of the screen. Consider the point, Q, in the object and its projection, P, on the screen. If the normal to the screen from the eye is EST, where S is its intersection with the screen and QT is the perpendicular from Q to the normal, then triangles ESP and ETQ are similar since SP and TQ are both perpendicular to the normal. In the figure, D and d are the distances of the eye from T and the screen respectively, and H and h the distances of Q and its projection from the normal respectively. Then:

$$\frac{h}{H} = \frac{d}{D}$$

$$\therefore\ h = H\frac{d}{D}$$

In this relationship, one can see the property which is commonplace. The further away an object is the smaller it appears. Although the relationship is simple mathematically speaking, perspective projections have not been used in the normal run of engineering design on account of being difficult to carry out using geometric constructions on paper. Furthermore, there is the disadvantage that one cannot take measurements from it. Instead, the parallel projection is used in which the line from each point in the object is drawn parallel to the normal to the screen (or paper) instead of to the eye as shown in Figure 14.2.

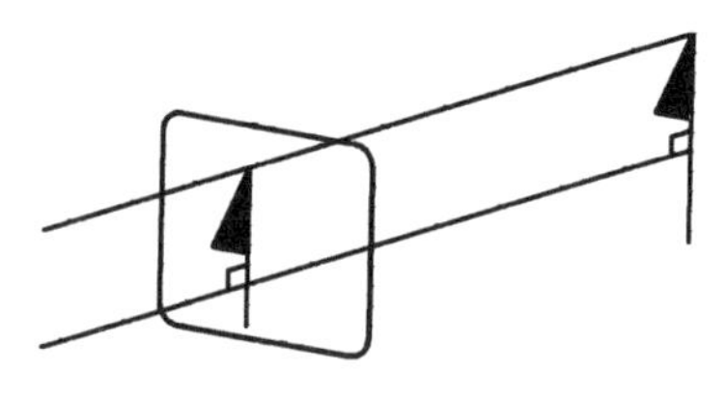

Figure 14.2 Parallel projection

Then $h = H$ and measurements can be taken, remembering that edges are scaled by the cosine of the angle they make with the screen. The projection still aids visualisation particularly if the important surfaces are tilted with respect to the paper or screen.

With a computer to do the work, perspective projections can be offered as a display option in 3D software. Surprisingly, a perspective projection is quite difficult to set up. A look at the formula will show why. In order to calculate the projection the software needs to know the distance of the eye from the screen and the position of the eye in the coordinate system of the model. The distance from the screen is usually about 500 mm and a constant can be used, but the position of the eye in relation to the model can be anywhere and must be chosen by the user. The problem is no different from trying to take a photograph in a crowded room where anyone close to the camera completely blocks the view. There is nothing to stop you locating the eye point 0.001 mm from a flat surface or even inside a solid, resulting in a blank screen or a view which is completely mystifying. Or you could position yourself with your back to the model looking out into empty space! Until there are display processors which will do instantaneous perspectives with hidden line removal it is wise not to try to locate the eye point by trial and error. Instead, calculate roughly or estimate some coordinates which are a reasonable distance from the object. You will also need to choose a suitable point in the model towards which to look (a "to" point) or to estimate a suitable direction vector. A grid on the $Z = 0$ plane of the model can be very helpful in estimating the coordinates of the two points.

It is quite easy to get distorted or highly exaggerated perspective views. The factors governing this are the position of the eye point and the angle of view - that is, the angle the edges of the screen make with the eye point. To understand the effect of the eye point position, consider a long plank H cm wide, L cm long which is oriented with its length perpendicular to the screen but tilted slightly upward so as to make its far end visible as in Figure 14.3. Let its near end be D from the eye point.

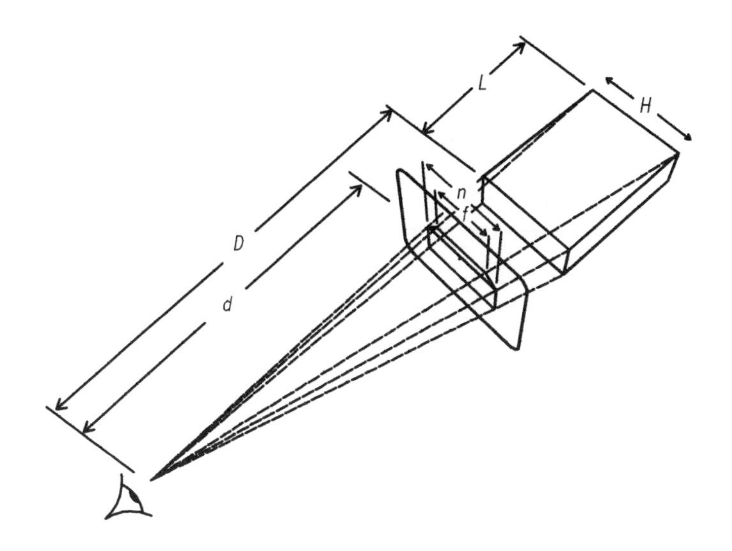

Figure 14.3 Exaggerated perspective

Then:

$$\text{Projection of near end } n = H\frac{d}{D}$$

$$\text{Projection of far end } f = H\frac{d}{(D+L)}$$

where d is the distance of the eye point from the screen. Dividing, we have:

$$\frac{n}{f} = \frac{(D+L)}{D}$$

$$= 1 + \frac{L}{D}$$

The ratio, n/f, of the projected sizes of the two ends varies with the distance of the plank from the eye. It has a minimum value of unity when the near end is a long way from the eye compared with the length of the plank. The nearer the plank the larger is the ratio. There is no upper limit to L/D since there is no lower limit to the distance, D , between the eye and the plank. The closer the end of the plank, the larger the ratio and the more exaggerated is the perspective effect and the more distorted the picture will seem. One way of reducing the distortion is to reduce the size of the screen so that the near end of the plank at the edge of the screen is cut off as shown in the figure. Alternatively, the projected view can be scaled up which is equivalent to reducing the size of the screen. The size of the screen can be represented as an angle - the angle of view, or the angle the eye moves through from one side of the screen to the other. Thus, the wider the angle of view, the larger the screen and the more exaggerated the perspective. The very same effect is obtained in photography when wide angle lenses are used.

A more detailed treatment of projections can be found in Reference (9).

Windows or viewports

The software can greatly help to make the best use of the screen by providing more than one window on to the drawing as shown in Figure 14.4. This has always been the case in 3D systems because of the need for more than one projection to aid visualisation and provide access to the three-dimensional model. It has taken 2D systems longer to realise the great value of having an overall view of the entire drawing on the screen at the same time as a highly magnified view of the section currently being worked on. Where multiple windows are provided there is a number of further facilities to consider. It is useful for the user to be able to decide the relative size and shape of the windows he has defined. He may wish to have a long narrow window at the top of the screen for the overall view and the rest of the screen for the detailed view, or he may want a large window for the overall view and a small one for the detail. Then it is very useful to see where the detailed view is located in the overall drawing. The cursor position should be shown in all windows. Also, any changes made in one window should be shown in all the other windows.

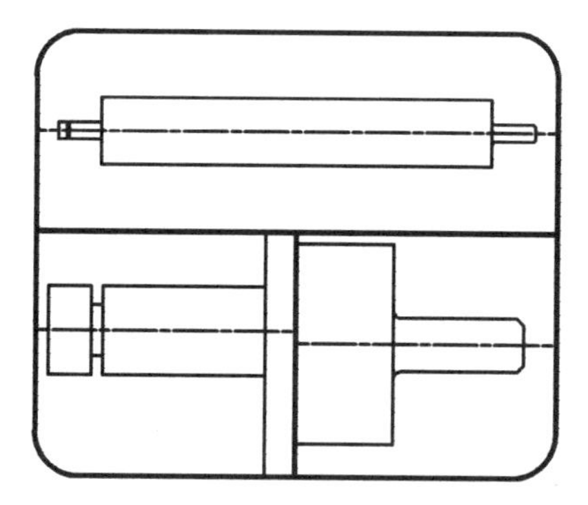

Figure 14.4 Windows and viewports

A term which is sometimes used in this context is *viewport*: standing for a particular frame on the screen through which the drawing is viewed. The term *window* is best reserved for the set of parameters - scale, view plane, eye point etc - which determine how the drawing will be projected through the viewport. In some software, any set of values for these parameters can be saved under an identifier for use later. When recalled they can be assigned to a particular viewport. A particular arrangement of viewports with their window assignments can also sometimes be saved under an identifier for recall later. In this case the identifier is called a screen identifier. Both facilities allow rapid switching from one view to another.

PLOTTING FACILITIES

Presenting the drawing on paper has much in common with presenting it on the screen. It must be scaled and if it is a 3D drawing it must be projected. The scaling will follow usual practice and be the stated scale of the drawing, a round number to aid interpretation. The scale is the ratio between the real-world distances of the design as held in the CAD drawing and the actual distances drawn on the paper. It is worth noting that certain things are not subject to scaling. These are the purely graphical items such as text, arrowheads and other draughting symbols. The software will need to distinguish draughting symbols from symbols which are part of the design. This can be tricky if the user is allowed to generate his own draughting symbols using the same mechanism for generating symbols in the design.

There is a significant difference between 2D and 3D systems in the way drawings are prepared for plotting. 2D systems actually mimic pencil on paper. What the user generates on the screen is very closely related to the marks on paper even though there is considerably more data present. Plotting out is simply a matter of handing the CAD drawing file to the plotting routines. The scale should be part of the drawing data. 3D designs, on the other hand, are usually created in a large empty space. Before they can be plotted out, a series of projections have to be defined and then arranged on a defined 2D area representing the plotted sheet along with annotation as required.

In both kinds of system, some provision has to be made to say which graphical entities are drawn by which pen. Various approaches are used. One is to link layers to pens, another is to allow each entity to take a parameter indicating the pen. But what happens when a particular pen breaks down? It must be possible to reassign the pens in the plotter driver program after the drawing has been completed. Another very valuable feature is software line thickening in which a line is traversed several times, each pass slightly offset from the next. The advantage is that one can generate a wide range of line thicknesses with ball points or just one liquid ink pen. The liquid ink pen is kept on the move without any idle periods when it can dry up. Another way of solving the problem of pens drying up is for the software to start a pen off with a few strokes outside the margin of the drawing.

Plotting is a slow process so it is essential to be able to plot without interrupting the use of the workstations. To do this, the software should maintain a plot queue into which drawings to be plotted are placed. Where the plotter is running unattended it takes its next plot from the beginning of the queue. In order to cope with pens clogging, paper running out etc there should be ways of manipulating the queue: altering the sequence, replotting a drawing, cancelling a drawing or temporarily suspending a drawing. For unattended operation there must be ways of separating drawings for different types of media, perhaps by putting them in different queues, so that the plotter only does the drawings intended for the medium currently loaded into it. A desirable feature for roll plotters which is not widely available at the moment is intelligent arrangement of drawings on the roll. Usually, drawings are placed with the long side along the roll solely to accommodate A0 drawings. This means that large amounts of paper are wasted for every other size.

There is a significant part of every drawing which never changes. This is the border. If pre-printed sheets hand mounted on the plotter are used then the border can be omitted from the drawing file, but if not then it has to be included. To reduce the size of the drawing file it is best held outside the file and merged with it as a reference type of symbol. Alternatively, merging the two can be done by the plotter driver.

EXERCISE

What is the difference between the transformations described in the previous chapter and zooming and moving the image on the screen?

Chapter 15 Solid modelling

It was probably the realisation that the 3D wire-frame representation of things, superior as it was to paper and pencil, was still limited in its ability to represent objects that provided the motivation to produce solid modellers. There was an obvious problem in visualising the object. You cannot tell which line is behind and which in front, and even when you know it is sometimes difficult to make one's mind see it that way round, particularly when viewing something from below. Besides the visualisation problem there is nothing in a wire-frame model to say unambiguously which part of space is inside the object and which outside, and it is possible to draw objects which can never physically exist such as the well-known one shown in Figure 15.1.

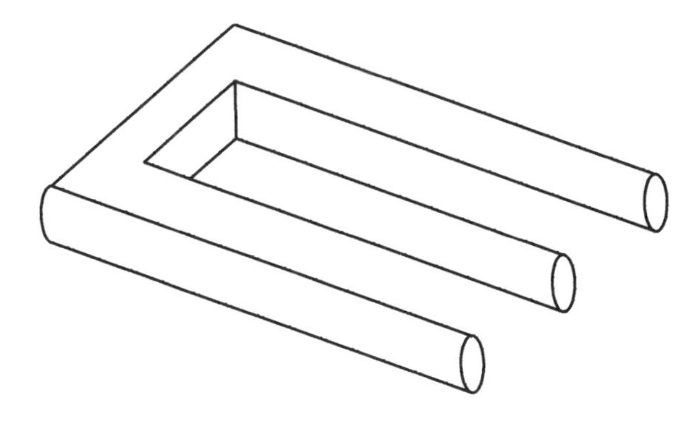

Figure 15.1 An impossible "solid"

This means that the software cannot calculate the volume of anything other than a basic shape without being given additional information. Surface modellers were not much better since they can model impossible objects with gaps in their surfaces allowing you to go inside the "solid" without passing through its surface. The user has to ensure that the portions of surface meet without gaps or let the software interpolate bits of surface to cover over the gaps.

Seeing that the world is made of solid objects the case for representing designs as solids almost goes without saying. The trouble is that the computational problems are so very much greater than those for lines or even surfaces. Furthermore, what language can be provided for the user to specify the geometry he wants? These two factors will continue to influence both the use and the development of solid modellers for some time to come. Engineering has managed for a century or more without a totally rigorous, unambiguous means of defining a solid object. Engineering drawings have always been interpreted, when the time has come to make the thing, by an intelligent human being. As intelligence of this degree is beyond the current power of computers, solid modellers cannot interpret a conventional drawing so another means of defining an object has had to be developed. Fortunately, what has emerged is quite closely related to many of the operations performed in the workshop.

SOLID MODELLING METHOD

The method consists of firstly creating basic "primitive" shapes and then combining them. One or more of the following methods are provided for generating the primitives:

1. Parameterised primitives such as cuboids, spheres, cylinders and toroids. The exact shape can be determined by a few dimensions in each case, e.g. height, length and width for a cuboid or radius for a spherc. In addition, the coordinates of the centre or a particular vertex must be given to position the primitive.

2. Linear sweeps of closed lines (see Figure 15.2). A closed line, or cross-section, which can be composed of straight or curved segments is moved a specified distance along a straight line. The resulting shape is a prism whose surface is defined by all the positions of the cross-section as it moves. The ends of the prism will be planes parallel to the plane of the line which must lie in a plane, otherwise the shape of the faces at the ends is undefined. This method of generating a shape is similar to a milling operation or an extrusion in plastics moulding.

3. Rotational sweeps of a line (see Figure 15.3). A centre line is specified and a generating line, composed of straight or curved segments, is swept round a circle centred on the centre line. The operation is similar to turning, the centre line being like the axis of a lathe. The circle will be in a plane perpendicular to the centre line. The generating line is usually prohibited from crossing the centre line since it is not clear in that case what shape is intended. If it is permissible for the generating line to stop short of the centre line then the shape is bounded by planes at its ends perpendicular to the centre line. Otherwise it is a requirement that the generating line ends on the centre line. Normally, the generating line is rotated through a full revolution

to produce a full solid of revolution but it is quite possible to allow an angle to be specified so that a sector of the full solid is generated instead.

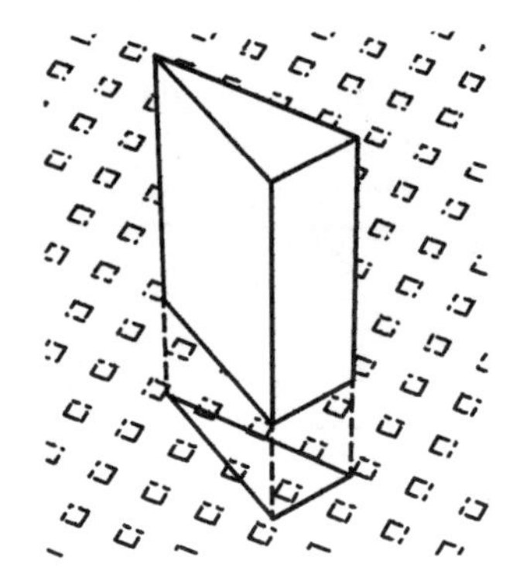

Figure 15.2 A linear sweep primitive

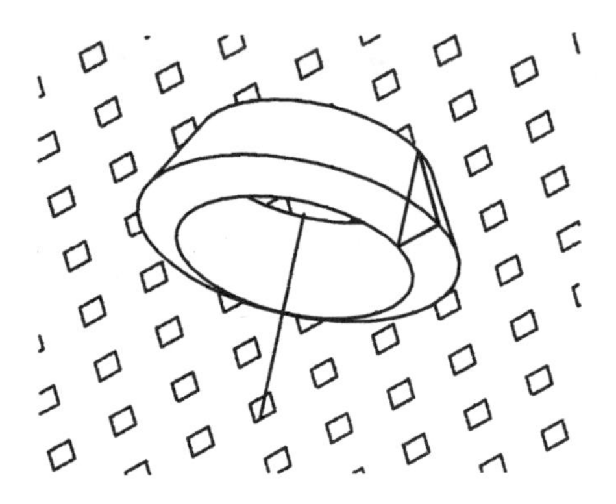

Figure 15.3 A circular sweep primitive

4. Sweeps along any chosen space curve (see Figure 15.4). This is an extension of method 2 in which the straight line is replaced by any line in space composed of curved and straight segments, usually called a *spine*. Additional data is required to control the generation of the solid. Firstly, a point related to the cross-section must be supplied in order to position it on the spine. The spine then passes through that point. Secondly, there has to be some convention or other means to control the rotation of the cross-section within its plane as it slides along the spine. The plane of the cross-section will be held perpendicular to the spine at all times so that it will swing about as it slides.

This means that there is no constant frame of reference for the cross-section as it sits in its sliding and rotating plane. One way is to provide a second spine and an additional vector in the plane of the cross-section. The two points where the spines pass through the plane then define a direction with which the vector can be aligned.

A condition which has to be avoided is when successive positions of the cross-section intersect each other due to a bend in the spine. This is similar to when the generating line in a rotational sweep crosses the centre line and it is no longer clear what shape is intended.

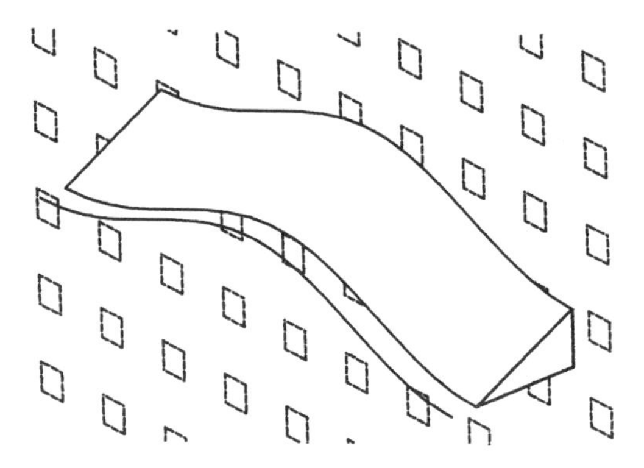

Figure 15.4 A space curve sweep primitive

5. A series of cross-sections. The software interpolates a surface over them as shown in Figure 15.5. Adjacent cross-sections are subdivided into equal numbers of equidistant points and corresponding points in each cross-section are joined by straight lines. The correspondence between the points has to be indicated by marking two points which are to correspond. The remaining points are taken in sequence round the outlines. If the two points are not opposite each other then twisting effects are produced. Such a surface is known as a ruled surface. The solid is bounded by the surface and the planes of the two end cross-sections.

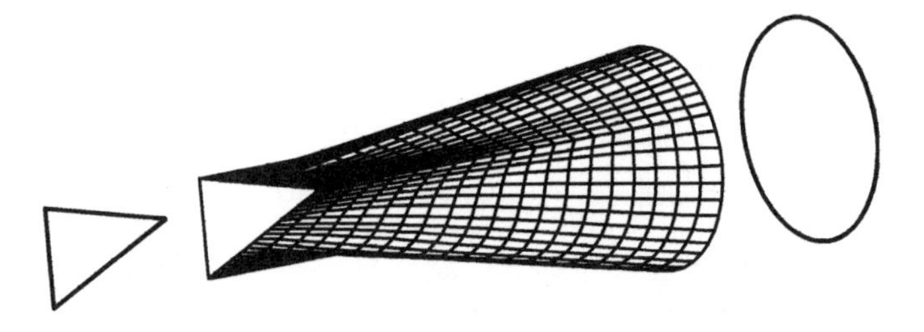

Figure 15.5 Use of sections

6. Two orthogonal views. This is only really suitable for defining planar faces. It is also used for defining pipes where only the centre line and a diameter are needed. Two views provide a convenient way of defining the centre line.

7. A set of surfaces enclosing a volume. The designer must ensure that the edges touch all the way round or allow the software to interpolate a surface across the gaps.

8. Two parallel copies of a surface offset from each other. The edges of the copies are joined with a surface by the modelling software.

9. A basket of space curves as shown in Figure 15.6. The modeller interpolates bounding surfaces in the meshes of the basket. The designer must ensure that the curves precisely intersect where they cross each other, otherwise the basket does not define a single bounding surface.

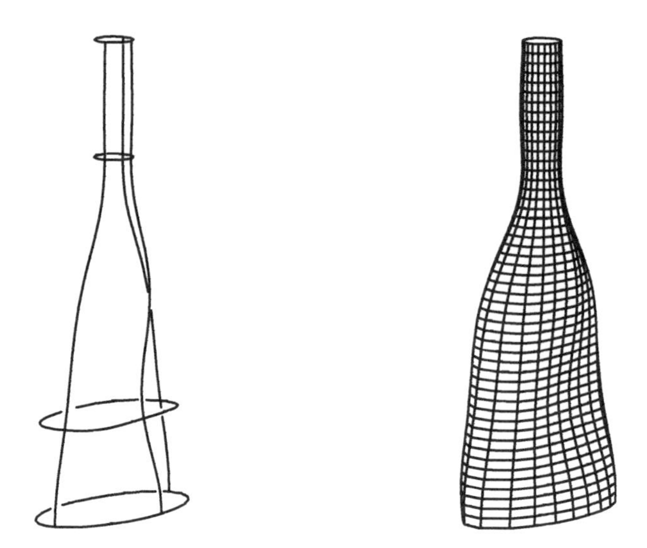

Figure 15.6 Use of a basket of curves

Usually, the shapes generated by these basic functions are insufficient to model the complex and intricate shapes of normal engineering components. It is necessary to combine a number of them. Modellers provide three ways of combining solid shapes which, although mathematical in origin, do relate to workshop practice. They are referred to as *Boolean operations* because they correspond with the operations of Boolean algebra. In all cases the two primitive solids should overlap so that the whole of space is divided into four regions as shown in Figure 15.7:

1. Space which is in neither of them

2. Space which is in both of them

3. Space which is in the first solid only

4. Space which is in the second solid only

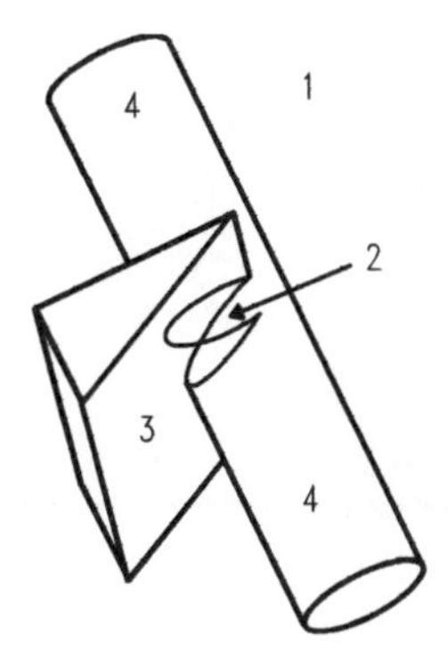

Figure 15.7 Two overlapping primitives

The three possible operations select different combinations of regions 2, 3 and 4.

The operations are:

Union The resulting solid is composed of space which is in either or both solids, i.e. regions 2, 3 and 4 as shown in Figure 15.8. The operation bears some relation to screwing or gluing objects together in the workshop.

Intersection The resulting solid is composed only of the region which is in both at once, i.e. region 2 only as shown in Figure 15.9.

Subtraction The resulting solid is composed of the region which is only in one of the solids depending on which one is being subtracted. If the first solid is being subtracted from the second the result consists of region 4, the space in the second solid only, as shown in Figure 15.10. The operation is analogous to cutting. In planning a modelling exercise, portions can be cut away by designing a solid to act as a cutter in a subtraction operation. The cutter solid will have an active portion with carefully defined geometry where it intersects the solid to be cut. The rest of the cutter which passes through empty space can have any convenient shape desired as it will not affect the shape of the result.

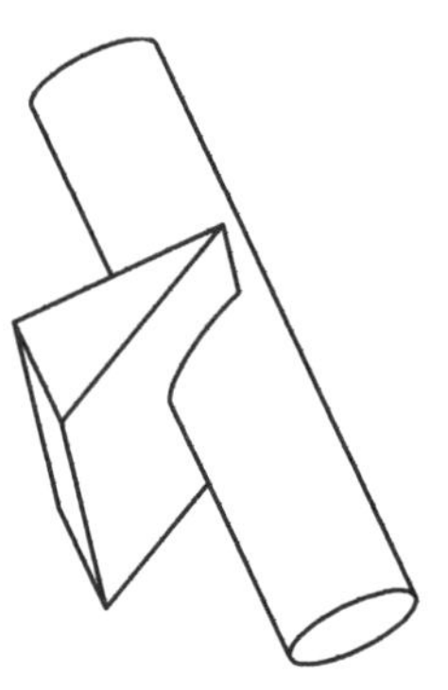

Figure 15.8 Union

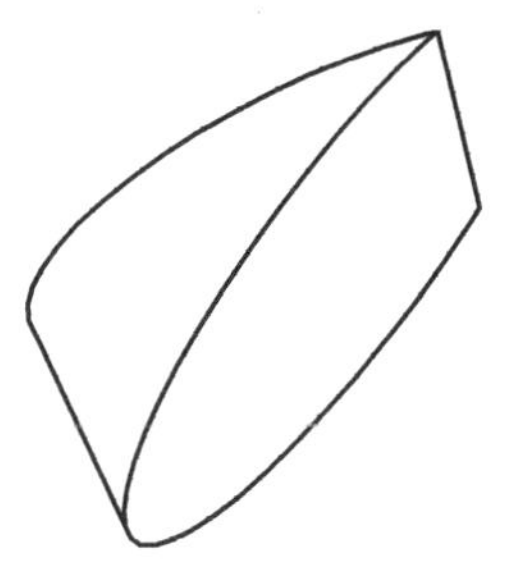

Figure 15.9 Intersection

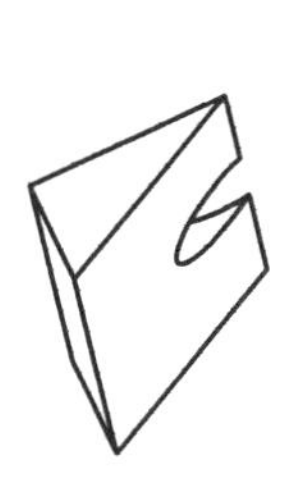

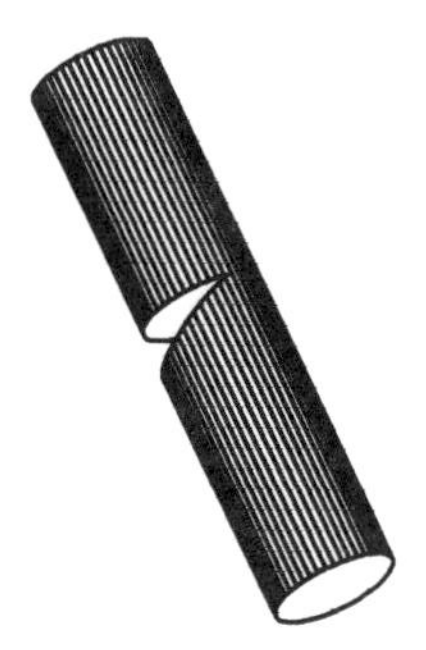

Figure 15.10 Subraction

Another way of modifying a basic shape is to split it with a plane or a surface and perhaps delete one of the solids produced.

REPRESENTING THE SOLID IN THE COMPUTER

Two principal ways are used to represent the solid in the computer which affect the performance and facilities made available.

Firstly, one can store data describing the vertices, edges and faces. This will take the form of the coordinates and other geometric data for each vertex, edge and face plus the connections between them, such as which vertex is at the junction of which edges. The method is known as the *boundary representation* or *Brep*. Alternatively, one can simply store the sequence of basic shapes and operators used to produce the solid. This is known as the *constructive solid geometry* or *CSG* method.

The two methods are analogous to either noting the result after evaluating a formula or instead noting the formula and the values to put into it. The Brep method stores the geometry of the actual solid but has no record of how it was produced, which means that if one wants to make an alteration one has to go back to the beginning and start again. The CSG method records how the solid was produced but an evaluation of the sequence of operations has to be performed before one knows anything about the actual shape. Clearly, there are advantages to both. If you are using a Brep modeller then it is a good idea to have some means of recording, editing and playing back the sequence used to produce it. With a CSG modeller you may want to keep the result of an evaluation for quickly seeing the solid it has produced. Some modellers do use both in fact. Further treatment of the techniques involved can be found in References (20) and (35).

FACETED MODELS

There are two variants of Brep modellers according to how the surfaces lying between the edges are held. These surfaces are portions of the primitives cut off by the Boolean operations along their intersections with each other. If the operations which generate the shapes produce surfaces which are too complex to store mathematically, a very good alternative is to approximate every curved surface by a large number of small plane faces. The data structures become simpler as does the software for handling all subsequent operations, since every solid is bounded by a surface consisting only of plane faces. The amount of computation and hence the response time increases with the number of faces. Since the accuracy of the approximation also increases with the number of faces, the user can choose between fast but approximate calculations and slower but more accurate ones. Often, the accuracy is controlled by a simple parameter entered by the user so that he can choose precisely how much he wants to trade accuracy for speed. Such models are called faceted models. Figure 15.11 shows a faceted model of a cylinder produced for different degrees of approximation.

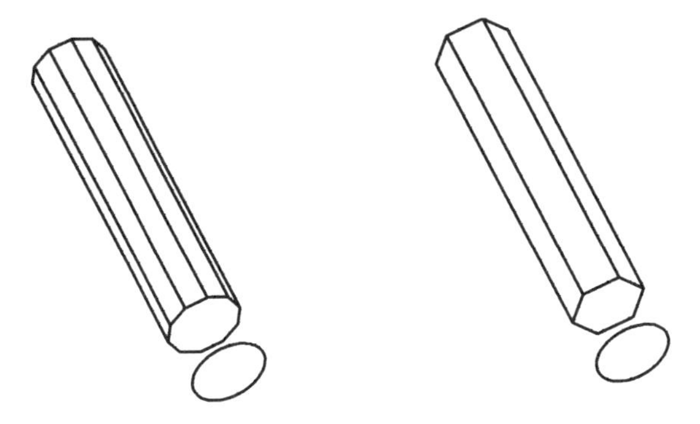

Figure 15.11 Faceted model of a cylinder

Usually, faceted models work well but occasionally the method breaks down badly. For example, if a thin walled pipe is modelled by subtracting one faceted cylinder from another, the corners of the facets on the inner "cylinder", shown in Figure 15.12, can break through the outer "cylinder" producing a pipe with slots along it as shown in Figure 15.13!

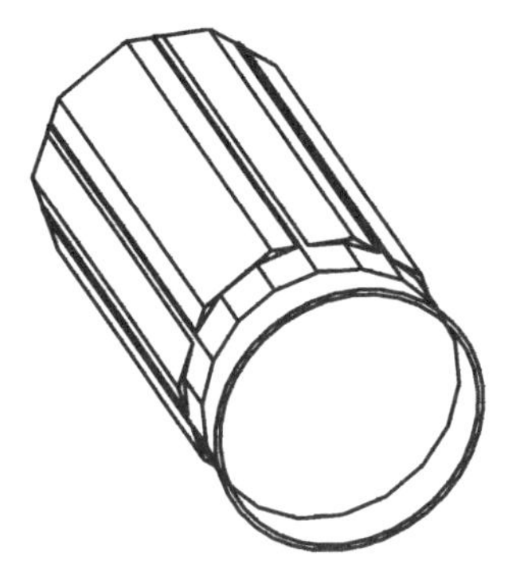

Figure 15.12 Failure in faceted models - inner cylinder

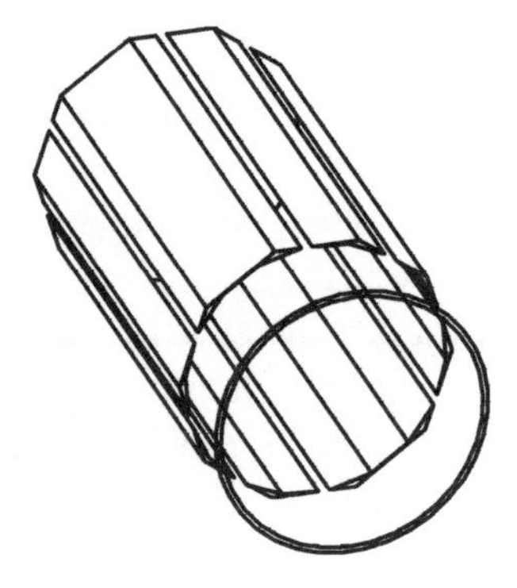

Figure 15.13 Failure in faceted models - result of subtraction

Another problem introduced by faceted models occurs if the designer has constructed a solid model to start with. This is a good way of doing design as solid modelling is easy and produces a complete representation of the object. However, when the designer comes to produce engineering drawings by projecting appropriate views of it and dimensioning them, he finds that the circles and arcs have become multi-sided polygons which do not have a diameter or a centre. Furthermore, some accuracy has been lost due to the approximations made by the modeller.

Faceting is often useful in visualising a model even when all the intermediate steps have been calculated precisely. The shape of a curved surface is best visualised in a line drawing if the surface has been faceted and the edges of the facets drawn. Somehow, the edges of the facets give a very good idea of the curvature.

MODELLING TECHNIQUE

When performing a Boolean operation between two primitives the program has to find all the faces of the two which intersect each other and, for each pair of intersecting faces, calculate the curve of intersection. It then has to create the two new faces which join along that line. The task requires extensive processing which makes solid modelling slower than normal interactive wire-frame geometry construction. In addition, the program has to find faces and edges which actually coincide and deal with these separately as special cases since the solution for the points of intersection could produce mathematically embarrassing results such as infinite numbers. Just occasionally a modeller will miss one of these special cases and break down with a fundamental arithmetic error. The modeller is usually unable to indicate which surfaces or edges gave the problem and the user has to guess where, in all the complex three-dimensional relationships, they are. Having located them, he then has to make small adjustments to the geometry to move the surfaces apart slightly. All this takes time and, unfortunately, the nature of engineering design and of CAD encourages the construction of geometric items which do coincide exactly. Fortunately, the occurrence is rare in current modellers. Nevertheless, it is advisable to build up the model in stages adding one Boolean operation at a time and evaluating each time. If you are only using modelling for visualisation you may be able to do without Booleans altogether, provided you can create the shapes without subtractions or intersections. Hidden line removal usually works adequately on overlapping "un-Booled" solids, producing an effect closely approximating to what would be obtained with a Boolean union. Certain inner edges where the solids intersect will be missing but their absence will usually not be noticeable. Volume calculations on such objects will, of course, be incorrect.

Planning a modelling exercise is very similar to planning the construction of something in the workshop. There are two approaches: making it in several parts which are then fixed together or starting with a single large block and carving it down. As no physical material is consumed in the process the difference between the two approaches is largely a matter of judgement on the relative convenience of each in particular circumstances. Coincident surfaces may be more difficult to find in a series of cuts crossing each other as the various solids defining the subtractions will be for the most part invisible.

If your modeller will allow it, you may find it convenient to make each component solid separately in its own coordinate system before combining them. If this is done then some thought will have to be given to ways of defining the orientation and positioning of each component in relation to the others. Once again, locating objects on a machine tool bed or positioning them relative to each other is a common workshop problem. In solid modelling, however, there are no genuinely solid surfaces to hold one against the other. Positioning and orienting will involve careful choice of the datum in each component. Creating each component separately allows different faceting tolerances to be applied to each item when using a faceted modeller and can also reduce computation time.

As mentioned earlier, ensure that you have a record of the sequence of operations used if you are not using a CSG modeller.

Feature-based modellers

Much of what is done in a CAD system is directed towards defining geometry and therefore uses the language and concepts of geometry. Although geometry is the underlying language in design, engineers use a higher level of concepts or entities based on commonly occurring three-dimensional forms or features. Examples of such entities are a hole - a circle bounding a void in solid material - or a boss - a prismatic projection from a solid object. The basic geometric operations give the full degree of flexibility but since so much of design uses this repertoire of engineering features the ability to specify a solid in terms of features is extremely valuable. A feature has a particular shape and topology but its dimensions vary from one instance to another. The solid is therefore designed by calling up each feature in turn, supplying parameters to specify its key dimensions and then indicating its location and orientation. A powerful facility is to be able to change the parameters after the feature has been added to the model.

In the conceptual phase of design, attention is given more to form and topology than to precise dimension - hence the use of sketching in the process. Topology is largely a matter of spacial relationships and therefore of solids. A feature-based solid modeller with a gridded sketching facility and the ability to alter the parameters of the components of the final solid shape, whether primitives or features, after it has been created becomes a useful tool in the conceptual phase of design. Rapid colour shading to assess spatial relationships adds to its usefulness.

VISUALISING OBJECTS

The problems of setting up a perspective projection are discussed in "Screen handling and output facilities" on page 133. Solid modellers usually provide hidden line removal in perspective and parallel projections. Because the result is a line drawing, curved surfaces are not shown very well. Their shapes can only be visualised from their edges, unless they are faceted when the edges of facets give a good idea of the shape. In many cases this is not a serious drawback where curved surfaces do not play a large part in the model, such as in machinery composed of flat faces and cylinders.

The shape of curved surfaces is best appreciated from the way light is reflected from them and so a colour-shading module is often provided. The appearance of a surface depends on quite a number of factors and colour-shading programs vary considerably according to how many of these factors are taken into account or are under the control of the user. The factors are:

1. The intensity of light reflected from each point on the surface

2. To what extent the reflection is specular, i.e. concentrated in a particular direction

3. The colour of the surface

4. The number of light sources

5. The size and shape of light sources

6. The colour, position and intensity of each source

7. The shadows cast by one part of the model on another as a result of each source

8. The refraction, colouration, attenuation and diffusion of light passing through the model

9. The texture of each surface - grainy, pitted, lumpy etc

A basic colour-shading program is most likely to take into account the first three factors using a single diffuse source of light, i.e. each surface is illuminated by light of the same intensity coming from the same direction. The curvature of surfaces will be seen from decreasing brightness as the surface inclines away from the direction of the light. The lack of shadow casting will not be too noticeable in simple engineering objects except when the inside of a hollow object is visible, where it will be unexpectedly bright as if there was a light inside it. If the colour of each surface is to be controlled
by the user then there must be some way for him to identify each surface of the model individually. Often, he is only allowed to assign a colour to a whole solid at once.

More advanced colour-shading programs allow for several point sources of light of different colours and will do shadow casting. Going further and including all the other factors leads to programs which take many hours to execute, thus placing a high price on simulated realism.

To make good use of colour shading you will need a high resolution screen with an adequate range of colours. Not only will the screen and software have to reproduce the colours you wish to give the surfaces but they will have to reproduce a good range of the darker shades of those colours to give the effect of shading. The software may generate the darker shades by putting in black pixels randomly. Besides a good colour screen you will need a good quality colour printer. Photographic transparency film, having been under continuous development for more years than any other colour-recording technique, is likely to give the best quality of all. If you do it by putting a camera in front of the screen, all the usual care required in photography needs to be taken, such as firm support for the camera, a darkened room to avoid reflections in the glass of the screen, use of the right film and correct exposure. Special colour printers using electrostatic or thermal processes are available and are gradually improving in quality but transparencies will always be the best way to capture the brilliance and colour contrast of a colour screen.

APPLICATIONS OF SOLID MODELLING

Solid modelling often seems to be a solution looking for a problem to solve. Very good visualisations can still be produced by getting an artist to render up a wire line perspective projection, surface modellers have been able to generate closed volumes and calculate their volumes and centroid positions etc and in any case engineering has managed very well for a century or more with 2D drawings. However, it is in the saving of labour and the greater generality of application that solid modellers excel.

There are three distinct areas where solid modelling can be used to advantage:

- Visualisation
- Analysis of volume properties
- Clash detection

Visualisation

Solid modelling has been used very extensively for visualisation. A good perspective drawing or colour-shaded picture is very valuable in sales proposals for those types of business where every contract involves special design to the customer's requirements. Drawings produced by CAD have a more definite appearance than an "artist's impression". Building the model may take longer than a technical illustrator takes to do a single picture but, once built, many views from different positions can be generated without further labour. If the product proposed is made of an assembly of stock components the model for the proposal can often be built quite quickly by assembling copies of models of the components from a library which has already been built up.

Visualisation is not only needed in sales, it is often needed for communication within a company itself: for supporting proposals to higher management or assisting installation or manufacturing departments (particularly for complex pipework). Solid modelling should be considered as an alternative to industrial scale models of process plant for deciding if a proposed design is convenient to operate and maintain. It is possible to locate the eye anywhere inside a solid model. This cannot be done in an industrial model.

Analysis

The advantage that solid modellers have over surface modellers is the vastly greater range of complex shapes possible, including solids with empty space within them. As the result is a true solid and not just a volume enclosed more or less by surfaces the accuracy of the calculated volume, surface area, centroid position etc is assured.

Clash detection

Because solid modellers can show the shape and position of and calculate the volume of any portion of one solid that overlaps another solid they are uniquely suited to finding and correcting clashes in a design.

EXERCISE

What are the strengths and weaknesses of solid modelling compared with surface modelling?

Chapter 16 Numerical control program generation

Numerically controlled (NC) machine tools arise as a result of putting accurate digital measuring devices on the movements of machine tools and using electronically controlled motors to position them. Numbers coded into electronic signals can then position spindle and workpiece precisely without a human operator. If these signals are recorded on tape with other codes to activate the cutter motor and any auxiliary devices, clamping devices etc then the entire cycle of the machine can be controlled by a tape. Which codes achieve what is decided by the designer of the machine tool control circuits and can be different for different models of machine. The wide use of NC machine tools led to the development of languages for specifying the program of the cycle independently of the particular codes required for a particular machine. Using these languages, NC machine tools are programmed like a computer, the work being done by a part programmer who studies the drawings of the component and, using his knowledge of machining practice and the machine tool in particular, writes a program for the machine in the language. The program is compiled and the result is a cutter location file which is still independent of the particular codes of the machine tool. This file is then converted by a program written specifically for the machine, known as a post-processor, into the particular codes it requires. Figure 16.1 illustrates the process. For a fuller treatment of NC programming see Reference (11).

Various part programming languages have been devised from time to time. The one most widely used is APT. It is now incorporated into an international standard (Reference (12)). Numerous sub-sets of the language have been developed for various purposes.

The cutter location file is literally a sequence of cutting tool positions held in binary numbers. The logical structure of the file has been standardised (References (13) and (14)) but the actual physical format will vary from one part programming system to another. For example, a floating-point number may occupy 4 or 8 bytes depending on the software producing it. As it is in binary it cannot be printed out as such although it is frequently processed on to the screen or printer via a utility program. The important difference between cutter location code and part program code, besides one being in binary and the other in text, is that the part program works at a much higher level. In other words, where the part

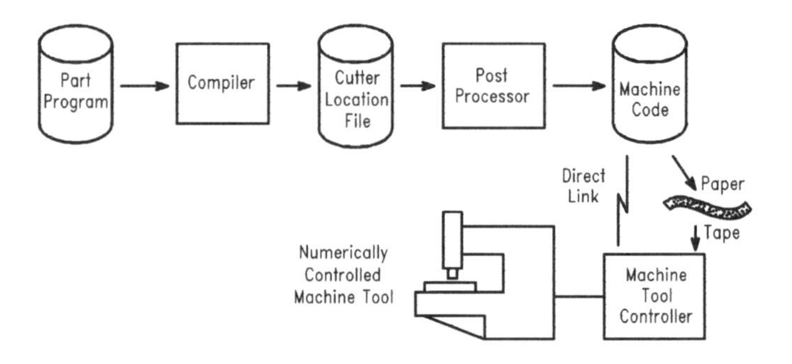

Figure 16.1 Numerically controlled machine tool programming

programme may have one statement requiring that a pocket be machined out, the corresponding cutter location code will contain all the positions of the cutter as it goes round and round the pocket.

Machine code is the actual sequence of codes read by the machine tool controller. Like the cutter location code, it has become standardised also, the form eventually adopted being the "word address format" (see References (15) and (16)). Despite the standardisation, machines still vary over what they require from the machine code in, for example, the number of decimal places used for coordinates or the way of doing circular interpolation. Machine tool suppliers may invent new uses requiring new codes or there may be no suitable codes for, say, pocketing routines in the machine tool controller. It is for this reason that the cutter location code has to be processed by a post-processor which is specific to a particular model of machine controller. Because of the many machine controllers produced it is always a problem finding post-processors to serve the particular controllers in use by a company, and so a CAM system is often judged by the extent of its post-processor library. This will be discussed in more detail later under the topic of selecting CAM software.

With the reducing cost of processing power there is little reason why the machine tool controller should not accept a higher level of language and some controllers have been produced which will accept a cutter location file directly as input. A new standard (known as BCL) has been produced for this purpose (see Reference (17)).

LINKING CAD TO CAM

If the drawing of the part is digitally encoded because it has been produced on a CAD system there is the opportunity to convert automatically the CAD drawing into a part program. You might think that as a result there would be no need for a part programmer as all the information is present. Unfortunately, it is not as simple as that.

A machine tool program includes the following types of information:

1. The path the cutting tool follows when cutting.

2. The path the cutting tool follows before and after cutting which must avoid the clamps holding the workpiece in position.

3. The speed with which the cutting tool moves which must be chosen in relation to the quality of the finished item, the load the machine tool can stand and the life expectancy of the cutting tool.

4. The sequence of passes where the complete job requires several separate passes.

5. Activation of auxiliary devices.

The only type of information the CAD drawing can possibly supply is the first in the list above: the path the cutting tool follows when cutting. The other four types have less to do with the actual shape of the component, being determined more by the characteristics of the particular machine to be used.

Even then the path of the tool when cutting is not determined entirely by the shape and dimensions of the component as given in the CAD drawing. Firstly, the component is rarely produced in one operation. There can be several intermediate operations as metal is progressively removed. Each stage may involve repositioning the workpiece on the machine bed, for example to machine the underside. None of the intermediate stages are drawn by the designer yet each produces a separate physical object for which a cutter path has to be generated. The intermediate shapes are determined by purely production considerations such as the balance between throughput and quality and the shape of the blank at the start.

Secondly, no product has a single set of dimensions. The drawing does not represent a single unique object. It represents a quantity of almost but not exactly identical objects: a population of objects. That is what tolerances are for: they specify how much variation is permissible within the population. But the geometry in a CAD drawing is that of one unique object corresponding to the nominal dimensions. Although tolerances appear in the drawing they are only text and are not held in the data in the same way that the coordinates of the vertices are held, and are not therefore usable by any automatic part programming system. All that NC path generation software can read from the drawing is the geometry of this unique nominal object. The designer might as well print "or thereabouts" after each dimension as far as NC path generation is concerned.

Once programmed, the NC path cuts an object upon which the actual variations and errors of the cutting process between one object and another will be superimposed to produce the population of production components. But will this actual variation meet what the designer has specified? Only if a human being has read the drawing and interpreted the tolerances, using his knowledge of the machine tool, into suitable feed rates and cutting depths. But what if the cutting errors come out equally spaced either side of the nominal dimension when the designer has specified tolerances which prohibit any divergence below it? An automatically generated path will be no good and human interpretation and intervention is needed. An important factor in CAD/CAM implementation is ensuring designers produce CAD models with dimensions at their mean values.

Lastly, machining experience is needed to determine the order of cuts. Many possible sequences of cuts can turn a particular blank into the desired result but the optimum sequence requires experience. Designing it automatically is beyond current practical computing techniques.

It is clear that if automatic NC path generation is to be employed then designs must be done specially for it. Anyone who talks glibly of just feeding a CAD drawing into NC path generation software is either ignorant of the realities of machining or trying to sell you something!

CAD FACILITIES FOR NC

Having noted the things a CAD drawing cannot contribute to a part program we must nevertheless admit that the geometry accounts for a large part of the information in the part program and this the CAD system can supply. Because of its numerical nature it is more open to human error in transcription so there is a distinct advantage in reading the geometry automatically from the CAD system. Thus, although the CAD drawing cannot be automatically turned into a finished part program, it can provide a large part of the data needed. The other information must come from people with manufacturing and machining experience.

The simplest provision a CAD system can make for NC program generation is just to output to a file the geometry of any line or surface selected by the user. The file is then input to a separate specialised NC program. Such an approach makes sense as the program will often be generated in a separate department which will prefer its own choice of software. However, many systems like to offer a complete package and this makes sense as the kind of facilities useful in defining a cutting path have much in common with those for defining geometry.

The main function an integrated CAD/CAM package provides is determining the path the cutting tool must follow in order to produce the shape defined in the CAD drawing. Having determined the path the software will usually draw it on the screen to allow the user to assess it visually. When satisfactory it will then be output to a file either in the form of one of the NC programming languages or of a cutter location file.

In determining the cutter path, several things have to be taken into account. It is the path of a particular point in the cutting tool since the tool is a solid object. As the tool moves, it sweeps out a closed volume in space and the shape produced is the result of subtracting this volume from the material being cut. To generate the path, one has to work backwards from the shape presented by the CAD design to the path which will produce it. Generally, the tool has a cylindrical or spherical shape and so the calculation is largely a matter of offsetting the tool centre from the desired surface by the tool radius. Performing these calculations is well suited to CAD software and provides a valuable facility, particularly where complex three-dimensional curves are involved. Having said this, there are still problems which require human intelligence to solve when the surface being cut is so concave that it interferes with the tool.

Another important facility is in determining the pitch between successive passes through the material. To cut a surface, whether plane or curved, the tool has to follow a series of parallel paths. If the tool is cylindrical or spherical, a little cusp or scallop of material is left standing between adjacent paths corresponding to the intersection of the two circular cross-sections of each path. The software can calculate the pitch between paths for the desired height of scallop.

NC software can be classified according to the limitations set on the path which can be generated. 2D software generates a path on a specified plane and can be used for lathes, punches and flame cutting. 2½D software fully controls the path in two dimensions but only makes step changes in the third dimension. Full 3D three-axis software generates a path moving smoothly in three dimensions. Multi-axis software caters for machine tools in which the orientation of the tool axis can alter as well as its position. Using these machines with the extra control (up to six axes) it is possible to generate virtually any surface shape.

MEETING THE REQUIREMENTS OF MANUFACTURING

The facilities just described fit very nicely into a CAD system and ease the task of generating a tool path. But the NC program has to work in the manufacturing environment. Here, the main concern is to make the most efficient use of particular machine tools with particular resources and to maintain production when machines are taken out of service for maintenance. For the sake of minimising cycle times and to avoid the requirement for too much memory in the machine controller the part program must make as much use as it can of the particular subroutines built into the controller for performing such things as area clearance etc. But this can mean that a particular program is only efficient on a particular machine so that switching to another machine requires a new part program. Software which makes it easy to change machines will be an important consideration. Another aspect is the link with process planning systems. The NC program is closely tied in with production routings and a process planning program may need to store, record or register it.

It is therefore very important for production staff to evaluate the CAM end of a CAD/CAM package if one is being considered or to ensure that if the part programming department is going to use its own separate NC package or CAD/CAM system that the links between the Production and Design Department are satisfactory. Will the CAD system produce a language which the CAM system will accept? How will the file be transmitted between the systems? They could be on a local area network or share the same computer. Alternatively, a diskette or half-inch magnetic tape could be used. Above all, if the supplier cannot provide adequate post-processors, any investment in money or effort will be wasted as the aim of CAM is to cut metal efficiently using the available machine tools. The selection of CAD/CAM software will be discussed further in "Making the case" on page 225.

EXERCISE

Discuss the management issues in introducing CAD/CAM (not just CAD).

Chapter 17 Finite element analysis

AN APPROXIMATION TECHNIQUE

The finite element (FE) method is a technique for solving certain kinds of mathematical problem using approximations. As with all other occasions when approximations are made in mathematical calculations it cannot give a precisely correct answer. The answer is sufficiently correct for practical purposes provided the approximations have been made with an intelligent understanding of the physical characteristics of the problem.

The technique is used where a physical parameter, usually a displacement resulting from a collection of applied stresses, varies smoothly and continuously over the interior of a complicated shape and obtaining a precise mathematical function for the variation is either difficult or impossible. The method is closely analogous to approximating a curve using straight lines and suffers from similar drawbacks. The longer the line segments the further they depart from the true curve.

The shape is subdivided into a mesh of small elements with simple geometrical forms as shown in Figure 17.1. The assumption is then made that the variation of the displacement is a simple function, such as linear or quadratic, of the distance across the element. Using that assumption, the strain energy of the element is calculated in terms of the displacements occurring at its boundaries with its neighbours. The result is a set of relationships, one for each element, involving all the displacements at the edges of the elements. It is known that the displacements will distribute themselves so as to minimise the total strain energy of the shape and the condition for this is that the partial derivative of the total energy with respect to each displacement is zero. Forming these partial derivatives gives a set of simultaneous equations which can be solved. Thus, the problem has been converted into the solution of a large number of algebraic equations. The penalty is the loss of accuracy resulting from assuming a simple function for the distribution of displacement within each element. This loss of accuracy appears in two forms. Firstly, the displacement at any point inside each element is highly inaccurate because a gross

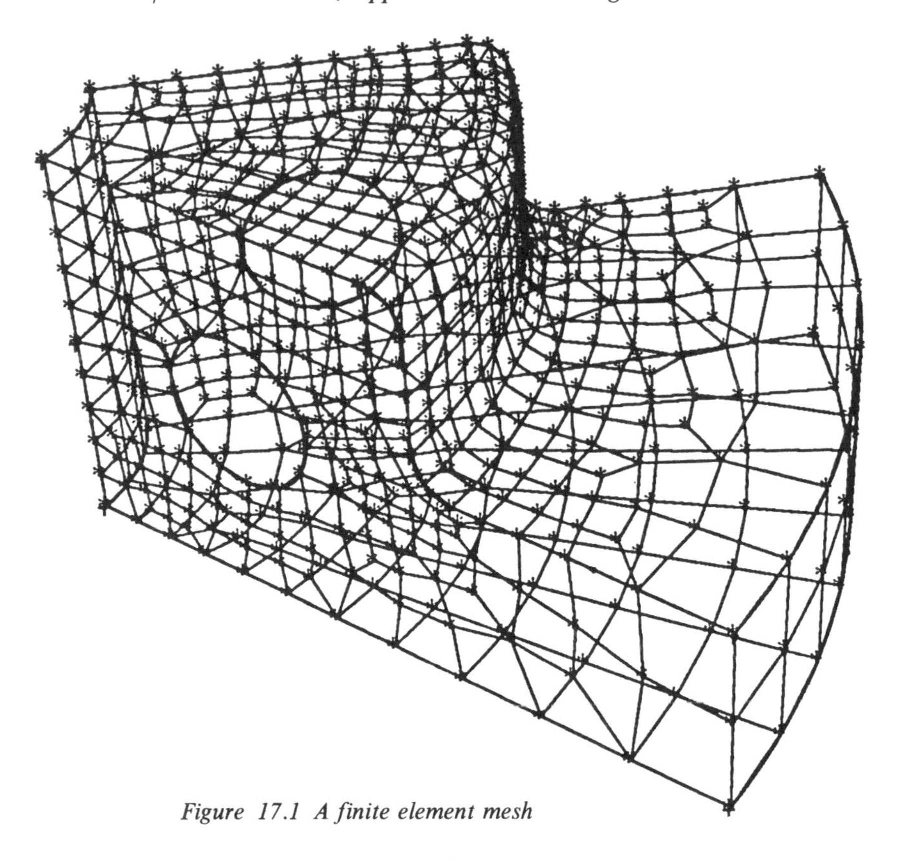

Figure 17.1 A finite element mesh

assumption has been made about its distribution inside. Secondly, even the displacements at the edges of the elements can be inaccurate.

ITS SUCCESSFUL APPLICATION

Having made such disparaging remarks, it must now be said that, in the hands of someone who knows how to handle its limitations, it is a valuable technique and essential to the solution of many problems. Anyway, engineering design, and even the applied physics on which it is based, makes considerable use of approximations. Many of the formulae used are based on approximations and much design data is empirical anyway.

The stress specialist using finite elements has to decide the size, and hence the number of elements to employ. In doing so he uses his judgement in striking a balance between having so many elements that the computing time is too long and having so few that the accuracy is impaired. He uses his experience of stressing to put in large elements at a low density where he thinks the stress will be low and small elements at high density where he thinks the stress will be high. When he receives the results he will check them against his experience to see if they "make sense".

THE ROLE OF CAD

The specification of the array or mesh of elements is a big chore as there can be very many of them in a complicated three-dimensional shape. For many years the mesh has been put into the analysis program as a large file of coordinates, thus providing opportunities for human error. Once the analysis was done the results were output as a table of displacements against nodes which is difficult to visualise and interpret. CAD software has provided useful aids in both creating the file of mesh coordinates and visualising the displacements. The three-dimensional shape is naturally best created using CAD software. In many cases it will have been created in the course of the design process anyway. Mesh generation then consists of specifying a particular element size and using the interactive graphics facilities to define the portion of the shape to be filled by it. The software can then calculate the coordinates of all the nodes. After the analysis the three-dimensional graphics visualisation facilities can show the displacements either as exaggerated displacements of the mesh or as coloured contours in a sectional view.

The CAD/CAM software thus fulfills the functions of pre-processor and post-processor for the analysis program. Some suppliers have gone further and provided an FE analysis program integral with the CAD/CAM system. Such a provision may not be necessary. The interface between the CAD/CAM system and the FE program is a well defined one and the functions of the two can be well separated. The analysis will be under the control of a specialist who will have his own preference for the analysis program.

Where a CAD/CAM system has no FE mesh generator it is still possible to provide very useful assistance if there is some kind of data extraction interface. For example, the author produced a simple FE interface in the space of about a day using a solid modeller and a data extraction program. The shape was first modelled and then a set of horizontal sections were taken at separations equal to the height of an element. The sections were presented to the stress analyst on the graphics screen who then used the cursor to position points where he wanted the vertices of the elements (nodes) to be. The coordinates of the points were then extracted and formatted into a file for the analysis program.

CONCLUSION

In conclusion, FE analysis is an aid to the specialist stress analyst who knows how to use the tool. CAD/CAM systems provide aids in using the tool by taking the chore out of entering mesh coordinates and by providing visualisation of the stress patterns. If a component breaks and someone is killed, the argument will not be about the impressive pictures that appeared on the screen but on the numbers that came out of the analysis and the stress expert's interpretation of them.

We finish with a cautionary tale about computer analysis. A designer used the moments that a CAD system printed out about a solid that he had modelled. On applying a little engineering common sense to them he decided that something was wrong and investigated further. He discovered that the moments the software was calculating were moments of inertia whereas he had been assuming they were bending moments. The software had done the calculation correctly but had not made it clear exactly what was being calculated. The day had been saved by engineering experience and common sense but had the designer taken the impressively precise figures as gospel without further thought the story would have been very different.

EXERCISE

Discuss the strengths and weaknesses of FE analysis. What management issue needs to be addressed in its adoption?

Chapter 18 Schematics and electrical and electronic drawing

SCHEMATICS

We now come to the area of work which makes the most productive use of CAD of all. Schematics, whether electrical ladder diagrams, power distribution diagrams, telecommunications circuits, electronic circuits, hydraulic systems, pneumatic systems or chemical processes, all consist of symbols selected from a limited repertoire, arranged on the drawing and joined up by lines which represent connections. Figure 18.1 shows an electronic circuit. The advantage of being able to pick a symbol from a library without having to draw it each time is obvious.

Other aids to fast drawing are also provided. With a snap grid, where every point indicated by the cursor is forced to the nearest intersection of an imaginary grid of lines, the draughtsman only has to position the cursor near a terminal on the symbol for the connection to snap on to it exactly. A facility which allows designated areas of the circuit to be moved from one place to another without breaking the connection lines is also valuable, especially if there is enough intelligence to put in any extra corners required to make the connection lines all horizontal and vertical again. Parameterised symbols or ones which can be built up as required out of smaller parts are essential. For instance, electrical connectors vary widely in the number of pins, transformers vary in the number of windings, and switches have different arrangements of contacts.

But CAD provides much more than aids to fast drawing. As with mechanical drawing, it is the availability of complex data structures underneath the visible drawing that make it so very much more than a paper analogue. While in mechanical drawing it is the numerical geometry and the representation of space and solid objects which make the difference, in schematics it is the large amount of descriptive data held and its cross-referencing which make the difference. Every component has a considerable amount of data specifying and describing it. By way of example, an electrical resistor could have the following:

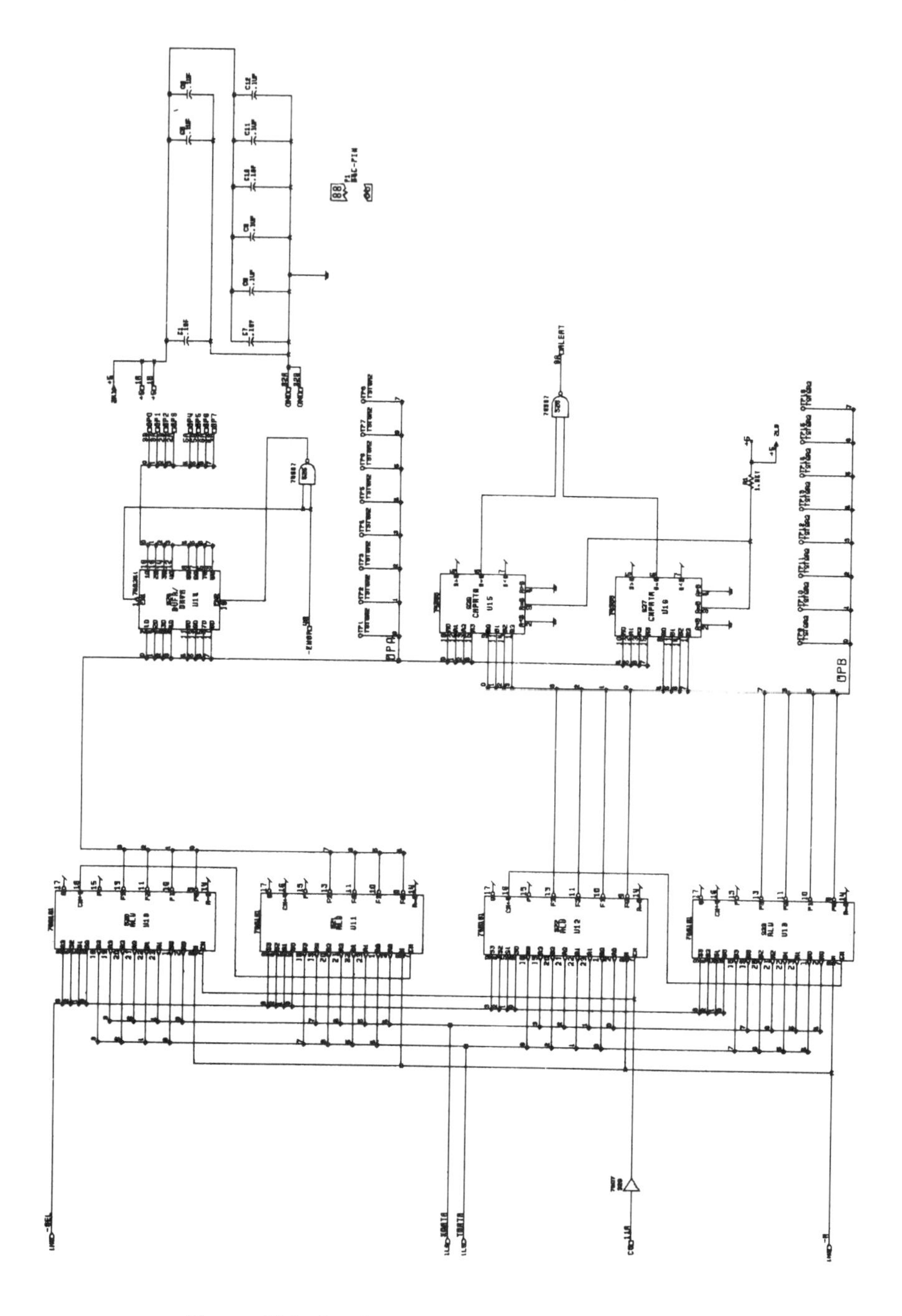

Figure 18.1 An electronic circuit schematic

- Schematic circuit reference
- Physical circuit reference
- Value
- Tolerance
- Rating
- Internal part number
- Supplier
- Supplier's part number

The first two items are unique to the particular physical item. The remaining items are generic to all resistors like it and are best held in a separate catalogue referenced by the internal part number. However, some of them need to be shown on the schematic to enable it to be interpreted.

Electronic circuits are manufactured as a printed circuit board (pcb) in which the connections between components are etched in a thin copper foil bonded to an insulating board, usually made of fibreglass sheet. Figure 18.2 shows the etched foil connections of a pcb. The components are mounted on the reverse side as shown in Figure 18.3.

The printed circuit is drawn on a separate sheet and is a two-dimensional drawing of the physical location and shape of each component. Suitable cross-references for identifying each component need to be maintained between the two drawings. Cross-referencing is complicated by the practice of packaging several separate circuit elements in one physical component. For example, a single integrated circuit chip may contain four electrically separate logic gates. The electronic circuit is designed and the schematic is drawn in terms of the individual circuit elements, a separate symbol being used for each element. Later on it is decided which circuit elements to group together into which component. Thus, several separate symbols in the schematic will need to be cross-referenced to just one component in the printed circuit. But the printed circuit references are not known until the schematic has been completed and passed on to those doing the pcb layout. Once the layout has been done the schematic must be updated with the printed circuit references. The physical shape of each component will also be a symbol for use in drawing the printed circuit layout. Thus, each circuit element will have two different symbols to represent it which should also be cross-referenced in the symbol libraries. Similar considerations apply to the other kinds of schematic such as hydraulic and pneumatic systems

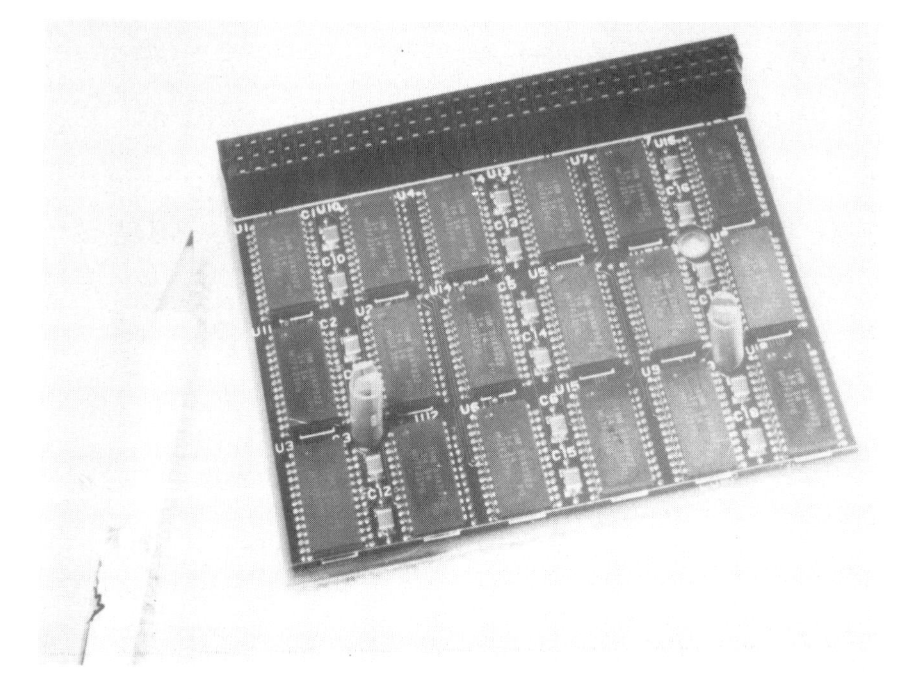

Figure 18.2 The etched connections of a printed circuit board

since they all represent the logical interconnections of things which have a physical implementation.

The foregoing indicates the need for quite an elaborate specialised database. Although it can be implemented by building suitable software round a general-purpose CAD system it is more efficient to design a CAD system specially for the application.

During the design of a circuit it is sometimes useful to be able to calculate its performance, particularly where bench measurements are difficult or unreliable. The software available for doing this requires details of the components and connections, which can be obtained from the schematic on the CAD system through a suitable interface. The results are easiest to interpret if shown graphically, which a CAD system is well suited to do.

PRINTED CIRCUIT BOARD DESIGN

Specialised software plays an even more important part in providing special aids to laying out pcbs which must conform to a number of constraints.

Figure 18.3 The components of a printed circuit board

The schematic specifies the connections that are to be made between the various pins on the various components. Laying out the board consists of deciding the positions of the components and the route each connection (known as a *track*) must take. The tracks are narrow strips etched in the foil on each surface of the board. One track can only cross under another by passing through the board (in a hole known as a *via*) to the other side. Both the width of and the gap between tracks are not allowed to be less than specified minimum values. A circular area of metal known as a *pad* must occur where the terminal of a component is soldered to the track. The layout should minimise the length of the connections.

The first step is determining the location of the components so as to minimise the lengths of the tracks. The process is aided by treating the connections as elastic bands joining the terminals and giving a display of the total length as different positions of the various components are tried.

Once the components have been positioned the tracks have to be routed. As this is a problem of logic with a small set of simple rules, software has been written to solve it, the programs being known as *auto-routers*. Although they save a considerable amount of human effort, auto-routers are not as clever as an experienced human being, so in practice they are used with discretion and in certain circumstances not at all. It can happen that the auto-router will fail after finding routes for about 95% of the tracks. The human operator then has to complete the job but he finds that it cannot be completed because the auto-router has not made very efficient use of the space available. For this reason, many users prefer to instruct the auto-router to stop earlier on in the job without trying to complete it because at that stage it is easier for the experienced human to complete. Certain kinds of circuit, such as telecommunications, radio frequency and audio frequency circuits, are subject to a further constraint which is that of avoiding a signal from one track passing across the gap to an adjacent track by capacitative coupling. Avoiding this problem requires considerable knowledge of the nature of the signals in the connections and designers prefer to do the routing manually.

If any manual routing has been done it is important to check that it has been done correctly. The software should keep a check of what connections have been routed as an indication of when the job is complete. Then it should be able to compare the tracks in the board layout with the schematic as a check that the correct connections have been made. It may also be necessary to check that the minimum spacing between adjacent tracks and the minimum widths of track have been followed.

In addition to the tracks, designers often like to leave unused areas of board filled with foil connected to ground as it provides useful shielding against capacitative coupling. The software should provide some means of drawing this in.

INTEGRATION WITH PRODUCTION

Once the board has been laid out the design needs to be implemented. The photo-etching process used to make pcbs allows the boards to be made directly from the design. To do this, an accurate full-scale transparency of the tracks is produced and the pattern etched directly from it using photo-chemical techniques. The pen plotter used for normal paper drawings is unsuitable to make the transparency because it is designed to draw lines whereas the pattern of a pcb consists of solid areas. A photo-plotter is used in which variously shaped patches of light produced by masks and a projection system are moved over a large photographic film.

A pcb does not only have a pattern of etched metal. It has holes for the vias and the component leads and it has conventional printing on it to identify components. The component identification is designed using conventional CAD methods. The holes can either be marked out and drilled by hand or made using a numerically controlled drill. If that is the case then a program for the drill can be produced automatically from the layout drawing in the CAD system.

In production the circuit boards are etched to produce the pattern of metal, printed with the component identification and then drilled. The components are then mounted in the board by pushing their wire leads through the holes provided. This is done by hand in many cases but can also be done by a machine which picks up a component, positions it accurately and pushes down with its leads through the holes. The positioning operation is numerically controlled like the drilling operation and again can be programmed directly from the board layout.

Not all connections in electronic equipment are made with printed circuits. The boards themselves have to be connected up and the units wired up in racks. Connections between boards are either made by hand or by automatic wire-wrap machines which position a wire-wrapping head over the terminals. The program for this can be obtained automatically from the schematic specifying the connections. For interconnections made by hand, which includes rack wiring, wiring lists are produced to tabulate the connections. These also can be extracted automatically from the schematic.

Finally, the schematic will contain a complete specification of all the components from which computer files can be extracted and passed to stock procurement and control systems.

It can be seen that electronic design currently provides the most complete opportunities for linking CAD to the downstream processes. Summarising, the data that can be extracted automatically from the CAD system comprises:

- Parts list
- Wiring list if manual wiring used
- Printed circuit board layout
- Printed circuit board drilling programs
- Automatic component insertion programs
- Automatic wire-wrapping programs

EXERCISE

List the facilities for designing schematics which a mechanical CAD system does not normally provide.

Chapter 19 CAD data exchange standards

This chapter and the two subsequent ones will tackle the increasingly important problems of data exchange and interchangeability of software. A standard of some kind is essential to any transfer of data from one program to another. The stream of binary digits constituting the data cannot be interpreted by the receiving program unless its format or coding scheme is known. It may seem trivial to say so but the only reason why standards are not an issue when programs in the same CAD/CAM system interchange data is that the people who write the programs are constrained by their management to agree a standard. When the programs are supplied by different companies there is no common management and standards become an issue and a problem. In this chapter we will consider a variety of standards which have been developed to help programs communicate with each other. When transfers of data take place between different computers in different places, hardware such as magnetic tape units and telephone lines are involved. We will describe some hardware standards in "Hardware data exchange standards" on page 203.

Software routines do not just pass data to each other: in many cases they call each other, that is, give each other work to do. A programmer writing a routine which calls another routine written by somebody else must know very precisely what process the subroutine will perform and in what form it will present the results. He also needs to know in what form the subroutine will expect the arguments and what it will do if it is unable to perform the processing for any reason. The communication between one routine and another is thus quite complex requiring careful documentation of the interface between them. Where a package of subroutines is supplied as a product by one company for other companies to buy and use, this is even more important. The interface is published and can become a standard.

The use of CAD/CAM has grown extensively in recent years. Competition has been strong and many different proprietary software packages, workstations and plotters have been produced and installed extensively. They are all using different data formats to represent the same thing: the definition, largely geometric, of a manufactured product. The case for better communication between systems is overwhelming yet it is hindered by what are in reality unenterprising defensive commercial policies going against the interests of the customer.

The many standards described here can be classified in a number of ways. There are three areas where standardisation has been attempted:

1. The CAD drawing or product definition itself.

2. Graphics packages of subroutines which provide standard display and user interaction facilities to the actual CAD/CAM program independent of the actual hardware used.

3. Device control languages providing standard commands to control the display, printer or plotter.

Some of the standards are simply data or file formats conforming to codes and syntax rules laid down by the specification while others are subroutine package specifications which define the parameters passed to the subroutines and the processes carried out.

Although all are standard in the sense that the definitions are public property, not all have the status of formal acceptance as official national or international standards. Some of them have been developed by particular companies which have then chosen to publish and maintain the specifications.

PRODUCT DEFINITION STANDARDS

The structure and organisation of the data in the computer memory which describes a CAD drawing will be complex. It is describing a wide variety of entities from text to solids and surfaces. It needs to ensure fast searching so that the nearest object to the cursor position can be found quickly. The performance and features of a CAD system depend on the ingenuity with which the data is organised and once the structure has been established it cannot be changed without creating problems for the user, who has to be able to take a drawing done some time ago and read it back into the system for further work. It is clear that those who write the software for CAD systems could not tolerate the data structure being dictated by others, and might even wish it to be kept secret. Unfortunately, companies send drawings to each other and they make their own individual choice of CAD system. It is therefore impossible for companies owning different CAD systems to exchange drawings without some way of converting between the various formats.

There are two basic approaches to the problem. One is to write a converter to go from each CAD system to every other. As conversion is required in both directions between a pair of systems, two converters are needed for each pair. If there were ten makes of CAD system then 2 x 10 x 9 = 180 converters would be needed. The other approach is to devise an independent format (called a *neutral* format) and write a pair of converters for each CAD system: a pre-processor to convert outgoing CAD data into the neutral format and a post-processor to convert incoming data from the neutral format. Our ten makes would then only require 20 converters.

Neutral formats have a number of advantages. Complex engineering products involve contributions from many subcontractors or even sub-subcontractors. Neutral formats enable precise descriptions of components in computer-readable form to be passed from one CAD system to another. The information transferred is of better quality and delays introduced by errors and misinterpretations are avoided. Then a company archiving its drawings in a neutral format allows itself the option of switching to a new CAD system when it needs to upgrade or finds its present CAD supplier unsatisfactory. The use of a neutral format for internal transfers of drawings allows different CAD systems to be used within the same company which in turn allows the optimum choice of system for each department. Another use of neutral formats which has yet to be exploited extensively is in parameterised drawing or special applications software generating engineering drawings directly in the neutral format. The advantage is that the special applications program will work with any CAD system without adaptation so that only one version of the software has to be written and supported.

The only disadvantage of neutral formats compared with special converters is that one cannot assume that once the neutral format has been devised every CAD system in the future will have data that can be converted into it. Special converters have the advantage that once written and tested you can be sure that they will work. Also, they may be more efficient through only needing to handle the entities common to both systems.

The need to exchange CAD drawings has led to a number of neutral formats being developed and a number of *de facto* standards arising. The principal neutral format in use currently is the Initial Graphics Exchange Specification (IGES) to which we will devote a separate chapter on account of its wide use. In this chapter we will describe some of the other neutral formats, standards and languages of significance.

SET (Normalisation Francaise Z68300)

SET stands for Standard d'Echange et de Transfert. Developed in the French aerospace industry it is an elegant and concise language deliberately aimed at avoiding the voluminous files generated by IGES.

As with all the other neutral formats it uses ASCII text. No line or record length is specified and spaces have no significance except within delimited text strings. The file is subdivided into blocks which are delimited by the "@" character at the start and sub-blocks delimited by the "#" character at the start. A third delimiter is the "," which is used to separate parameters. The first parameter of every block is an integer indicating the type of block and the second parameter is an integer sequence number. The remainder of the block consists of a sequence of parameters depending on the type of block it is. Some types of block or sub-block will be fixed length and others variable length according to type. The purpose of a block is to group together a number of sub-blocks, each of which describes a single entity. Parameters are FORTRAN reals (indicated by the presence of a decimal point), integers, text strings enclosed by single quotes, pointers consisting of a block sequence number preceded by "!" or dictionary parameters preceded by ":". Dictionary parameters are used to set values, such as colour, which are applicable to many types of entity. A single dictionary parameter setting can apply to a collection of sub-blocks in a block, thus avoiding duplication. An integer specifies the parameter. This is followed by a comma and the value it is to have.

The overall structure of the file can designate a series of assemblies marked out by start of assembly and end of assembly blocks. Sub-assemblies can be called within each assembly in the manner of a FORTRAN subroutine call.

VDA-FS (DIN 66301)

The aim of this was to aid the exchange of sculptured surfaces within the German automobile industry. It is a very simple standard with just the few entities necessary for the transmission of geometry. They are:

- A single 3D point
- A sequence of 3D points
- A parametric curve
- A parametric surface
- A named set of geometric entities

- A comment
- A file header
- A file terminator

The file is in plain text and most of the statements take a form similar to the following example which is the one for the single point:

```
P0001 = POINT/1.567,2.345,4.567
```

The item to the left of the "=" is a name. As can be seen the keywords are English.

VDA-IS

This is a subset of IGES recommended for use in the German automobile industry.

STEP and PDES

STEP, the Standard for the Exchange of Product Model Data, in draft at the time of writing, draws on the experience of previous data exchange standards and promises to become a really effective interchange standard for a wide variety of products. It aims to provide a neutral exchange mechanism capable of completely representing product definition data throughout the life cycle of a product. It will provide much more than exchange of geometry. Whereas its predecessors define just a file format, STEP provides a formal mechanism for describing the meaning of the data to be transmitted. A language, called Express, has been devised which uses the extensive data declaration methods employed in modern programming languages particularly Pascal. It is, in effect, a language for saying what particular attributes or parameters are required to specify an entity, what their characteristics are, what constraints apply and the relationships the entity has with other entities. There are also full procedural facilities for defining calculations.

The specification is divided into three layers:

- The application layer
- The logical layer
- The physical layer

The application layer carries definitions specific to particular applications such as Civil Engineering, Mechanical Engineering etc. It uses the data description language furnished by the logical layer. The actual file format is handled by the physical layer where a formal, unambiguous, machine-readable notation is used to specify the syntax of the file.

PDES, or Product Data Exchange Specification, has been separately funded by the United States defence industry. The work is being done in cooperation with that on STEP and will be an American contribution to STEP which is likely to emerge earlier.

Electronic Design Interchange Format (EDIF)

The Electronic Design Interchange Format is a data exchange standard devised in the USA principally for very large scale integrated circuit (VLSI) designs. Printed circuit board definitions will be added in due course.

Digital Mapping for Customers (DMC)

Digital Mapping for Customers is a standard for transmitting maps used by the Ordnance Survey. It is limited to straight lines and text.

The Autodesk DXF file

The DXF format was originally a format devised by the suppliers of the Autocad package to write out a description of the drawing as ASCII text for processing by other software such as stress analysis, bill of materials generators etc. It also allows drawings to be exchanged between Autocad on different types of computer. However, because Autocad is widely used and many separate software companies have written auxiliary programs which communicate with Autocad via DXF files, it is becoming a very useful neutral file format. Writing programs to analyse or generate DXF files is relatively easy.

The file is composed of pairs of lines, the first in the pair being an integer in FORTRAN I3 format, known as the group code, and the second a single item of data which can be one of three types: a text string, an integer or a floating-point number. The type can be deduced from the group code, 0-9 being strings, 10-59 floating-point numbers and 60-79 integers. The convenience for FORTRAN programs provided by this format will be apparent to those with experience of the language. Typically, entities will start with a group code of 0 followed by the name of the entity type and then various parameters as appropriate, each preceded by the particular group code.

The file overall comprises a header section giving parameters describing the drawing as a whole, a tables section giving tables of layer names, line types, text styles and views, a blocks section defining all the blocks (symbols) in the drawing, and an entities section listing all the entities in the drawing. The file concludes with an explicit end-of-file marker. The DXF input program in Autocad will allow the header, table and block sections to be omitted.

As a neutral file format it is limited to the repertoire of entities found in Autocad which is being extended as new issues appear. There is also a binary file format (DXB) which is more compact but not human-readable.

Intergraph Interchange File (IISF)

As its name indicates, IISF is another file format devised by a particular CAD vendor which has gained popularity on account of the strong market presence of its originator.

Computer Graphics Metafile (CGM)

The Computer Graphics Metafile standard has acquired some importance as a result of being included in the group of standards required for the Computer-Aided Acquisition and Logistics Support (CALS) requirements of the United States Department of Defense. It is a file format for expressing two-dimensional geometry using binary numbers (rather than human-readable character codes) and has some relationship to GKS in its concepts.

GRAPHICS PACKAGE STANDARDS

FORTRAN and other high-level languages allow programs to be divided into separate subroutines which are written, compiled and tested quite independently of each other. When they are all complete link editing takes place in which all the subroutines are joined together into a single program. This ability to write subroutines quite independently of each other means that a collection (or *library*) of subroutines can be written to serve a particular purpose and supplied as a separate product. All a user needs is the names, the operations they perform and the arguments required. Knowledge of the internal workings is not needed. If a library becomes widely used its interface definition takes on the nature of a standard. Furthermore, an interface could be defined and published quite independently of any actual subroutines. If the definition is rigorous, unambiguous and accepted as a standard then software writers can produce programs calling the subroutines knowing that they can be linked at any time to a library of subroutines conforming to the standard. At the same time, anybody who wishes can write a conforming subroutine library.

A truly universal standard should be usable in more than one programming language. This is possible because the most important part of the standard is not specific to any language, being concerned with defining the exact mathematical operations performed, the arguments that are used and the results returned. The part specific to the language will be the names of subroutines and the data types of the arguments. A version specific to a particular language is known as a *language binding*.

The justifications for a standard graphics package are program portability, so that a program written to work with the package can be transferred to any other computer using the package, and programmer portability, so that a programmer trained to use the package can write programs for any computer using the package.

A number of concepts have formed the basis of the current graphics package standards. For a discussion of these with particular reference to the CORE proposal see Reference (25). An important idea was to say that the program using the package had its own model of the application it was handling which was of no concern to the graphics package. The size of the model may be measured in millimetres, metres or kilometres. The job of the graphics package is to display whatever portion of the model the application program requires and to relay back any inputs made by the user, whether they are selected parts of the displayed image or keyboard input.

The application model will have a coordinate system, called the *world coordinate system.* The display, on the other hand, will use a separate two-dimensional coordinate system of its own for specifying a position on the screen. The application program passes portions of the model (lines etc) expressed in its world coordinates to the graphics package which then has the job of converting the geometry to the coordinate system of the display. The conversion involves a number of processes. Three-dimensional geometry has to be projected on to a viewing plane and the result has to be scaled to fit the display and perhaps rotated. The application program will only want a portion of the model displayed and will define this with a rectangular frame called a window. The package clips the lines off at the frame. The application program may want to use only a portion of the display for this part of the model so as to keep the rest of the display screen for other parts. It therefore defines a rectangular portion of the display called a viewport. To preserve independence from the actual display in use, the program defines the viewport in terms of *normalised device coordinates* in which the display screen is treated as one unit wide by one unit high. Since whatever appears in the window is what appears in the viewport, the scaling can be calculated by the ratio of their sizes. If they have different ratios of height to width, distortion will occur as there will be different scales in *X* and *Y* but the application programmer will be aware of this danger.

As the application program runs it will generate a list of things to display on the screen. To assist in the management of the display it is convenient to divide this list up into portions which the application can individually delete or alter. Such portions are called *segments*. As graphics often use repeated symbols it is useful to allow one segment to reference another segment rather than repeat the detailed display geometry it contains all over again. Such a facility leads to hierarchically segmented display data.

Input from the user may occur as keyboarded text and data, function button operation, screen position indications from a cursor or light pen, screen picks in which the user selects a graphical item displayed on the screen, and adjustment knobs (valuators) for panning and zooming etc. It is the job of the graphics package to pass this information to the application program in a way which is independent of the particular device in use. Screen picks will appear as segment identifiers. User input is classified into events initiated by the user and sampled devices such as screen cursor positions and valuators in which the program reads a position when it needs to.

Finally, the graphics package needs to handle the attributes of displayed items such as colour, line style and text style.

Graphics Kernel System (GKS ISO 7942)

The Graphics Kernel System has been developed specifically as a standard for the interaction between a graphics workstation and software generating pictures to implement a particular application. Several characteristics of the standard make it as universal as possible. It is not defined in terms of any particular programming language. The standard simply defines 209 graphics operations in terms of what they do and the parameters passed between them and the calling program. A GKS subroutine package can then be developed for each language. The workstation is not a particular hardware device but a set of idealised facilities such as a flat display surface with its own coordinate system, a keyboard, a pointing device which outputs some coordinates etc. The standard does not assume that the real workstation has any particular set of facilities but the programmer is given functions by which he can find out which facilities are actually available. The current version of GKS only handles 2D models although a 3D version is becoming available. A more detailed discussion with programming examples can be found in Reference (27).

The CORE System

The CORE System was developed in the USA by the Graphics Standard Planning Committee of the Association of Computing Machinery Graphics Special Interest Group in 1977 and offered as an alternative to GKS when the formulation of an international standard was being considered. It handles three-dimensional graphics. See Reference (26) and the papers accompanying it for a more detailed discussion.

PHIGS (Programmer's Hierarchical Interactive Graphics System)

The Programmer's Hierarchical Interactive Graphics System aims to provide more powerful functions than GKS: 3D graphical facilities in particular and a hierarchical segment structure.

DEVICE CONTROL LANGUAGES

The graphics packages convert the geometry of the application model into marks on the flat surface of the display or plotter. Device control languages arc standard ways of describing marks on a sheet of paper or display. The standards below include both subroutine packages and command languages.

The X Window System

The X Window System manages bit-mapped displays on computers and terminals networked together. In a CAD environment it provides a service to the CAD software with regard to displaying the drawing and handling input from the user via keyboard, mouse etc. It does not handle the precise geometry of the CAD model but only the user interface to that geometry, that is, its display on a screen, cursor control and screen menus. It is very flexible, allowing a program on any computer of the network to be accessed by any workstation on the network, and its code is independent of the type or model of the hardware used so that workstations of different vendors on the same network can interact freely. As a further encouragement to widespread application the standards documents and source code are in the public domain, made available for only a nominal fee.

The system was developed at the Massachusetts Institute of Technology (MIT) arising from a need in 1984 for a display environment for debugging distributed systems running under the UNIX operating system. This fairly narrow objective produced software which attracted interest and in 1985 it was made publicly available without a licence. A number of computer manufacturers then started to base products on it. The specification was still too narrow in its applicability for widespread use so to avoid the spawning of widely incompatible variations a major redesign was undertaken with assistance from the Digital Equipment Corporation Western Software Laboratory. Major software and hardware vendors gave public support and then in 1987, to ensure that MIT remained in control of the standard, the MIT X Consortium, funded by over 30 organisations including almost all the major US computer vendors, was set up to control subsequent development. It is becoming increasingly popular. Reference (23) gives a detailed description and Reference (24) a helpful introduction.

The system operates by placing a display server between the display and the rest of the system. All communication with the display passes through the server. Any application program wishing to interact with the display must make calls to the subroutines in the X Library which then communicate over the network to the display server using a protocol defined by the X System. Thus, X defines a subroutine library and a network protocol used by the library.

Calcomp plot calls

CAD system vendors like to be able to offer a variety of pen plotters to work with their system. Plotter vendors are the best people to write the code which actually issues the commands to their hardware so the practice has arisen in which the plotter vendor supplies a subroutine library with his plotter. Since Calcomp was an early plotter supplier, its set of subroutine calls has become a standard used by other plotter suppliers.

Tektronix graphics commands

The subroutine library that Tektronix provided for driving its displays (PLOT10) has set a standard for libraries of routines for driving graphics displays.

Hewlett Packard Graphics Language (HPGL)

Hewlett Packard devised a coding system to control its pen plotters. Instead of using binary codes it devised an ASCII coding system and the result is a language which is easy to interpret and generate and is being used outside its original application.

PostScript

PostScript is an advanced graphics language devised to control those graphics printers which can draw straight lines and curves, fill areas, generate text in a wide range of typefaces and reproduce digitised photographs (e.g. laser printers). It is therefore very much more powerful than a pen plotter control language which can only draw straight lines and (sometimes) curves. The language uses ASCII text and so can be read by humans. It is normally generated by a program, such as a word-processor or desk-top publishing program, but it is sufficiently easy to use for anyone with normal programming experience to generate graphics by writing a program.

PostScript has all the facilities of a modern programming language and, like a programming language, the order of the statements is significant. The statements place graphical items on a page, those placed later overlaying earlier ones where they overlap. The language uses the post-fix notation in conjunction with a stack. Operands are presented first and stacked. The operator comes last, takes its operands off the stack and leaves the result behind on top of the stack ready for the next operation. Powerful graphics operations are available such as filling closed areas, clipping, changing the intensity of the marks made, rotating and translating the current coordinate system, and manipulating type fonts. Further information is available in References (7) and (8). With the increasing use of laser printers, in which PostScript is used, more and more CAD systems will have an option to generate PostScript. This will integrate CAD with general graphics and desk-top publishing giving considerable aid to technical publishing departments.

Computer Graphics Interface (CGI)

CGI is a recently developed comprehensive standard for controlling graphics devices. It includes a subroutine package for use by a graphics package such as GKS or an application program and a set of codes for controlling the actual hardware. It uses some of the concepts of GKS and handles both vector data and pixel data.

CONCLUSION

As can be seen there are very many standards, each with its own mysterious acronym. Despite all the effort in making standards there seems remarkably little to show for it in the way of making things easier for the user or even the software developer. Several of the standards overlap. In product definition, STEP will eventually take over from IGES and promises to be a practical proposition. In graphics packages, there are clear deficiencies in GKS which may be overcome by subsequent versions but PHIGS is more likely to replace it. In device control, PostScript is already well established as a laser printer language and could also become a plotter language, particularly if laser printers replace pen and electrostatic plotters. PostScript does not have any user input features and therefore does not compete with X Windows or CGI. Of these two latter standards, there is such overlap between them that it is likely that CGI will be the loser. X has the advantages of handling networks, of being developed to meet a real need and of being adopted enthusiastically by device suppliers.

EXERCISE

Give the practical benefits of:

1. Product definition standards
2. Graphical package standards
3. Device control standards

Chapter 20 The IGES standard

The Initial Graphics Exchange Specification (IGES) is an international standard, Version 5.0 of which is now current. It is used widely with a fair degree of success. Early versions suffered from inadequate precision in the specification and an inadequate range of entities. Also, vendors adopted the sharp practice of carrying out only partial implementation in order to make claims that they had "IGES compatibility". Things are a little better at the time of writing and most users find they can establish successful transfers using it after a day or two of problem solving.

In an extensive test carried out in 1988 by the Motor Industry Research Association and the Organisation for Data Exchange and Tele-Transmission in Europe with the assistance of the Cadcam Data Exchange Technical Centre, six pairs of engineering companies from the automotive, aerospace and rail industries carried out drawing exchanges between four pairs of different CAD/CAM systems. Of 70 faults discovered, only one or two were attributable to inadequacies in IGES. About a third were connected with misinterpretations of the specification and the remaining two thirds or so were software bugs.

FILE FORMAT

IGES is a data format which uses ASCII - coded text in a sequential file of 80-character records. (There is also a binary form and a compressed ASCII form.) The specification refers to these records as lines. Each line has an identifier in columns 73-80. This consists of a serial number starting at 1 at the beginning of each section prefixed by a single letter. Considerable cross-referencing occurs using pointers which are the integer portion of the line identifier. IGES defines a record for its own purposes as a sequence of data terminated by a semicolon. Records are further subdivided into parameters by commas. There are six different sections to the file, the two principal ones being the fourth and fifth. The fifth, the parameter data section, contains the full descriptions of the entities and the fourth, the directory entry section, contains an index to the descriptions. In detail, the sections are as follows:

Flag Indicates if the binary or compressed ASCII form is being used. If the file is binary the first character is the ASCII letter "B" while if it is in compressed ASCII format there will be a "C" in column 73 of the first line.

Start — Free text prolog for humans to read. Lines have an "S" in column 73.

Global — The parameters needed to process the file. These include the product generating the file and the date of generation, the pre-processor and IGES versions, the formats for representing numbers, the largest coordinate value, the resolution, the scale, the units, drafting standard used etc. Lines have a "G" in column 73. Alternative record and parameter delimiter characters may be specified in the first two parameters of the record (see Table 20.1).

Directory entry (DE) — An index to the entities described in the parameter data section. All records have the same format and consist of two lines comprising 18 fields of eight characters, most of which are pointers to the various records describing the entity: in particular, the first line of its record in the parameter data section. The number of lines in the parameter data record for that entity is also recorded. Lines have a "D" in column 73 (see Table 20.2).

Parameter data (PD) — The description of the entities, consisting of a sequence of parameters for each entity. Parameters are separated by commas. The first parameter is the entity type number from which the meanings of the subsequent parameters are deduced. Each line contains a pointer in columns 66-72 back to the directory entry for the entity. A record delimiter is put at the end of the parameters. A comment may be added after this. Lines have a "P" in column 73.

Terminate — A single line giving the final line identifier of each section.

Table 20.1 Global Data record

Field	Description
1	Parameter delimiter character (default: ",")
2	Record delimiter character (default: ";")
3	Sender's identifier for the product
4	File name
5	ID for software producing the file
6	Version of IGES processor in sending system
7	Number of bits in integers
8	Maximum power of ten in a single-precision floating-point number in sending system
9	Number of significant digits in a single-precision floating-point number in sending system
10	Maximum power of ten in a double-precision floating-point number in sending system
11	Number of significant digits in a double-precision floating-point number in sending system
12	Receiver's identifier for the product
13	Model space scale (model unit/real world unit)
14	Unit (1=inch 2=mm 4=feet 5=miles 6=metres 7=kilometres etc)
15	Text string naming the units (to MIL12 or IEEE260)
16	Maximum number of different line thicknesses
17	Maximum line thickness
18	Time file generated (13Hyymmdd.hhnnss)
19	Smallest discernible distance
20	Upper limit to coordinate values
21	Name of person responsible for creating file
22	Name of organisation creating file
23	IGES Version number (6=Version 4)
24	Drafting Standard (1=ISO 2=AFNOR 3=ANSI 4=BSI 5=CSA 6=DIN 7=JIS)

IGES ENTITIES

IGES handles both geometric and non-geometric entities. The latter are further classified into annotation entities for notes and dimensions etc and structure entities for recording logical connections between entities. The geometric entities are listed in Table 20.3. A general principle is adopted in which an entity is defined in its own coordinate system before being positioned in the coordinate system of the drawing or model by a translation and a rotation. A separate entity, the transformation matrix (124), is provided for defining the translation and rotation. The copious data entity (106) provides for sequences of points in various forms. The parametric spline curve entity is a sequence of parametric cubic polynomials. The coefficients of the X, Y and Z polynomials of each segment of the curve

Table 20.2 Directory Entry record

Field	Description
1	Entity type
2	Pointer to start of entity's parameter data
3	If an entity defines this one's structure, a pointer (negated) to its DE
4	Line font pattern or a negated pointer to the DE of an entity defining one
5	Level (layer) of entity or negated pointer to DE of entity defining a list of layers
6	Pointer to DE of an entity defining the view
7	Pointer to DE of transformation matrix which positions the entity
8	If there is a label display associativity, points to its DE
9	Status: four pairs of digits indicating visible/blanked, dependency on other entities, usage and hierarchy
10	Line identifier
11	Entity type
12	Line thickness
13	Colour
14	Number of lines in the parameter data
15	Form number of entity
16	Unused
17	Unused
18	Entity label
19	Subscript to label
20	Line identifier

are recorded along with a parameter indicating whether the curve has slope continuity or both slope and curvature continuity at the break points.

In solid modelling only CSG entities (Nos 150-168, 180, 184, 430) are handled. Besides the commonly used primitives (150-168) and the Boolean tree (180) for combining primitives, there is a solid assembly (184) for combining primitives without Boolean operations, that is, when surface intersections are not computed. The annotation entities are listed in Table 20.4. The copious data entity is listed here on account of the forms it provides for centre lines and witness lines. The sectioned area supersedes the section form of the copious data.

The structure entities are listed in Table 20.5. The associativity definition entity and the associativity instance entity allow logical relationships between entities to be defined. An associativity is basically a collection of lists. Each list is termed a “class”. For example, there may be an association between a dimension and the entities it is dimensioning. To record this structure, an association definition entity is created to define the particular type, i.e. that there is to be an association comprising one

Table 20.3 IGES geometric entities

Entity no.	**Entity**
100	Circular arc
102	Composite curve
104	Conic arc
106	Copious data
108	Plane
110	Line
112	Parametric spline curve
114	Parametric spline surface
116	Point
118	Ruled surface
120	Surface of revolution
122	Tabulated cylinder
124	Transformation matrix
125	Flash
126	Rational B-spline curve
128	Rational B-spline surface
130	Offset curve
132	Connect point
134	Node
136	Finite element
138	Nodal displacement
140	Offset surface
142	Curve on a parametric surface
144	Trimmed parametric surface
146	Nodal results
148	Element results
150	Block
152	Right angular wedge
154	Right circular cylinder
156	Right circular cone frustrum
158	Sphere
160	Torus
162	Solid of revolution
164	Solid of linear extrusion
168	Ellipsoid
180	Boolean tree
184	Solid assembly
430	Solid instance

dimension linked to a list of the items dimensioned. Then the processor writes an association instance entity for every dimension it meets. The instance entity will comply with the format defined in the definition entity. Its directory entry would reference the definition entity and the list of items would consist of a list of pointers to the directory entries of the entities

Table 20.4 IGES annotation entities

Entity no.	Entity
106	Copious data
202	Angular dimension
206	Diameter dimension
208	Flag note
210	General label
212	General note
214	Leader (arrow)
216	Linear dimension
218	Ordinate dimension
220	Point dimension
222	Radius dimension
228	General symbol
230	Sectioned area

being dimensioned. A number of pre-defined associativity definition entities is specified in the standard.

The subfigure definition entity defines a symbol which is going to be used many times over. The network subfigure definition entity provides the additional facility of connection points in the symbol for schematics. Both list all the entities comprising the symbol which may include other symbols. The depth of the resultant nesting is recorded. The network entity also records whether an item in the symbol is for use in a physical layout or a schematic diagram. The actual instances of the symbols are recorded by the various subfigure instance entities which provide for rectangular and circular arrays as well as isolated occurrences.

The view entity specifies a viewing orientation of an object in space or a clipped projection of a 3D object into a 2D drawing. The drawing entity specifies the arrangement of various view entities in a 2D drawing together with annotation entities. The name property, drawing size property and drawing units property entities are also used in this connection.

The external reference entity allows a reference to an entity defined in another file. In the case where the file describes just a sub-assembly the entire file can be referenced.

The macro definition and instance entities allow new entities not covered by the specification to be defined using a programming language. The language includes expression evaluation and conditional execution. Figure 20.1 shows an IGES file and the corresponding model.

Table 20.5 IGES structure entities

Entity no.	Entity
0	Null
302	Associativity definition
304	Line font definition
306	Macro definition
308	Subfigure definition
310	Text font definition
312	Text display template
314	Colour definition
320	Network subfigure definition
322	Attribute table definition
402	Associativity instance
404	Drawing
406	Property
408	Singular subfigure instance
410	View
412	Rectangular array subfigure instance
414	Circular array subfigure instance
416	External reference
418	Nodal load/constraint
420	Network subfigure instance
422	Attribute table instance

MAKING IGES WORK

Experience up to the time of writing indicates that IGES is by no means an automatic trouble-free way of passing a model from one CAD system to any other and, considering the problems inherent in transferring CAD data, this is not surprising. Setting up a transfer should therefore be planned in this light. Problems can occur under the following categories:

1. Administration

2. Media

3. IGES processor

4. Incompatibilities

Administration

Since the transfer will most likely be taking place between two separate organisations at two geographically separated sites the normal principles of good liaison are essential in order to solve the problems as they occur. Each organisation will need to identify a responsible person or a contact for the exercise. Covering letters describing what is being sent will need to go with the tapes etc.

When a problem occurs its cause is most easily identified if the circumstances are as simple as possible, so the exercise should start with simple test data consisting of just one or two items of each type. Solve the problems on the test data before trying real data.

Where problems cannot be solved then some way of working round them will have to be agreed between the two parties. It may even be necessary to agree not to transfer certain items.

Media

The first problem to solve is ensuring that the data is sent on media which can be read at the other end. If standard half-inch magnetic tape is used then ensure that the density (bpi) is compatible and mark the tape with the blocking factor (lines/records per tape record/block). Other useful information, if relevant, is the number of logical tapes on the reel and the files per logical tape. If other types of media are in use then even more care is needed to ensure compatibility.

IGES processor

Since IGES has gone through a number of revisions up to its present Version, the versions at the two sites should be checked and the consequences considered if they do not match.

IGES files can be very large and therefore take a long time to process. This will not be the case with the test data which has been deliberately made small. If the processor takes too long then a bug in it will have caused it to hang in a loop. It is more likely to fail with some error messages. These should be recorded and given careful analysis. Immediate failure before processing any data will most likely be due to the wrong format in the file. The standard permits three different formats. On the other hand, the pre-processor in the sending system may be generating an erroneous format. An IGES file verification program, obtainable from the CAD/CAM Data Exchange Centre at Leeds University, will be of use here.

The slowness with which a processor handles an IGES file, whether on input or output, is partly due to the file being divided into separate directory and parameter sections. To read or write any entity requires access to both sections which will be situated at different parts of a long file. This causes the file access routines to continually search up and down the file. The slowness will be particularly apparent if a tape file is being accessed by the processor since the tape unit will have to keep on winding the tape back and forth. It is therefore advisable to get the processor to access a disk file. Transfer the file to or from tape as a separate operation. Even then the process can be further speeded up by setting up a separate I/O unit in the processor for the directory and parameter sections, instead of the more conventional single unit for the one file. Each unit maintains its own pointer into the file and the processor simply switches from one pointer to the other, instead of making the single pointer search up and down the file.

Incompatibilities

The pre-processor has to match up every entity in the sending system with an IGES entity and sometimes the selection of the right IGES entity is a matter of judgement. Similarly, the post-processor has to choose entities in the receiving system to match particular IGES entities. Where there is no hard and fast rule for matching IGES entities, a good processor will allow the user to indicate which one to choose by parameters supplied at the start of the conversion. The post-processor may well find IGES entities which it cannot convert but by adjustments to the pre-processor it may be possible to replace them with other entities which the post-processor can handle. In any case, those involved in planning the transfer should know what entities are used at each end.

Other aspects which can go wrong are the scaling and the position of the origin. The result is a drawing which appears blank until carefully studied, whereupon the received drawing is discovered to be a tiny dot in one corner of the screen! Scaling is a complex matter. Some systems will hold the geometry in real-world values and then convert to paper distances for the purpose of plotting while others may only ever hold paper distances. Text heights should remain independent of scaling.

Layers (or levels) can cause trouble. Different systems permit different numbers of layers and different design offices use layers in very different ways so that when a drawing arrives from a transfer everything is on the wrong layer. It can even result in the received drawing arriving on an invisible layer so that nothing can be seen at all. One way of dealing with problems of invisible drawings is to get a listing of the entities in the drawing.

Other areas which may need care and attention are line styles, symbols taken from a library and logical groupings.

Another source of problems occurs with polynomial curves. It is quite possible for the sending system to support a curve of a higher order than the receiving system can handle. The only way round the problem is to replace the single high order segment with a sequence of low order segments.

Many of the early problems with IGES were the result of deficiencies in the specification combined with poor implementations on the part of the CAD suppliers. Considerable work has been undertaken to improve the specification and the increasing prominence being given to CIM (computer integrated manufacturing) will create a climate in which CAD suppliers will be obliged to provide good quality IGES processors. IGES will eventually be superseded by STEP but so much experience has been accumulated and so many processors have been written now that IGES is likely to be of value for a number of years yet.

EXERCISES

1. On examining an IGES file which you have received from another CAD site you find many records with a "D" in column 73 and the numbers 320 or 420 in columns 1-8. What sort of drawing is it likely to be?

2. On another occasion you find many records with numbers between 150 and 184 in columns 1-8. What would you expect to find in the CAD model?

```
Illustration of IGES file                                              S      1
1H,,1H;,,8HCFAM20.1,14HIBM CATIA V3R1,8HIGES 3.0,32,75,6,75,15,,1.0,2,2HG      1
MM,10,1.0,13H910731.212811,0.10,119999.9988,,16HDASSAULT SYSTEMS,4,3;   G      2
     124       1       0       1       0       0       0       0       0D      1
     124       2       8       1       0                    AXS1       0D      2
     124       2       0       0       0       0       0       0   20000D      3
     124       0       0       1       0               TRANSFOR        0D      4
     402       3       0       1       0       0       0       0       0D      5
     402       2       8       1      16               PLANAR A        0D      6
     406       4       0       0       0       0       0       0       0D      7
     406       0       0       1    6004               PROP6004        0D      8
     124       5       0       0       0       0       0       0   20000D      9
     124       0       0       1       0               TRANSFOR        0D     10
     108       6       0       1       0       0       0       0   10201D     11
     108       2       8       1       0               CLIPPING        2D     12
     108       7       0       1       0       0       0       0   10201D     13
     108       2       8       1       0               CLIPPING        3D     14
     108       8       0       1       0       0       0       0   10201D     15
     108       2       8       1       0               CLIPPING        4D     16
     108       9       0       1       0       0       0       0   10201D     17
     108       2       8       1       0               CLIPPING        5D     18
     406      10       0       0       0       0       0       0   10000D     19
     406       0       0       1    6004               PROP6004        0D     20
     410      11       0       1       0       0       9       0   20101D     21
     410       2       8       1       0                     VU1       7D     22
     404      12       0       1       0       0       0       0     100D     23
     404       2       8       1       0                   DRAFT       8D     24
     406      13       0       1       0       0       0       0       0D     25
     406       2       8       1      16               PROP0016        9D     26
     124      14       0       0       0       0       0       0   20000D     27
     124       0       0       1       0               TRANSFOR        0D     28
     100      15       0       1       0       0      27       0   20000D     29
     100       2       8       1       0                    CRV2      11D     30
     110      16       0       1       0       0       0       0   20000D     31
     110       2       8       1       0                     LN1      12D     32
     110      17       0       1       0       0       0       0   20000D     33
     110       2       8       1       0                     LN2      13D     34
     124      18       0       1       0      21       3       0   20100D     35
     124       2       8       1       0                    DRAW      14D     36
     402      19       0       0       0       0       0       0       0D     37
     402       0       0       1       7                    SET1       0D     38
```

```
124,1.0,0.0,0.0,0.0,0.0,1.0,0.0,0.0,0.0,0.0,1.0,0.0,0,0;            1P      1
124,1.0,0.0,0.0,0.0,0.0,1.0,0.0,0.0,0.0,0.0,1.0,0.0,0,0;            3P      2
402,1,1,3,35,0,0;                                                   5P      3
406,1,1,0,0;                                                        7P      4
124,1.0,0.0,0.0,0.0,0.0,1.0,0.0,0.0,0.0,0.0,1.0,0.0,0,0;            9P      5
108,1.0,0.0,0.0,-1.0E+05,0,-1.0E+05,0.0,0.0,0.0,0,0;               11P      6
108,1.0,0.0,0.0,1.0E+05,0,1.0E+05,0.0,0.0,0.0,0,0;                 13P      7
108,0.0,1.0,0.0,1.0E+05,0,0.0,1.0E+05,0.0,0.0,0,0;                 15P      8
108,0.0,1.0,0.0,-1.0E+05,0,0.0,-1.0E+05,0.0,0.0,0,0;               17P      9
406,1,0,0,0;                                                       19P     10
410,1,1.0,11,15,13,17,0,0,1,5,2,7,19;                              21P     11
404,1,21,0.0,0.0,0,0,1,25;                                         23P     12
406,2,0.0,0.0,0,0;                                                 25P     13
124,1.0,0.0,0.0,0.0,0.0,1.0,0.0,0.0,0.0,0.0,1.0,0.0,0,0;           27P     14
100,0.0,0.0,0.0,50.0,0.0,50.0,-7.847704121E-14,0,0;                29P     15
110,3.552713679E-15,0.0,100.0,50.0,0.0,1.776356839E-14,0,0;        31P     16
110,-3.552713679E-15,0.0,100.0,-50.0,0.0,1.776356839E-14,0,0;      33P     17
124,1.0,0.0,0.0,0.0,0.0,1.0,0.0,0.0,0.0,0.0,1.0,0.0,1,5,0;         35P     18
402,3,29,31,33,0,0;                                                37P     19
S      1G      2D      38P      19                                  T      1
```

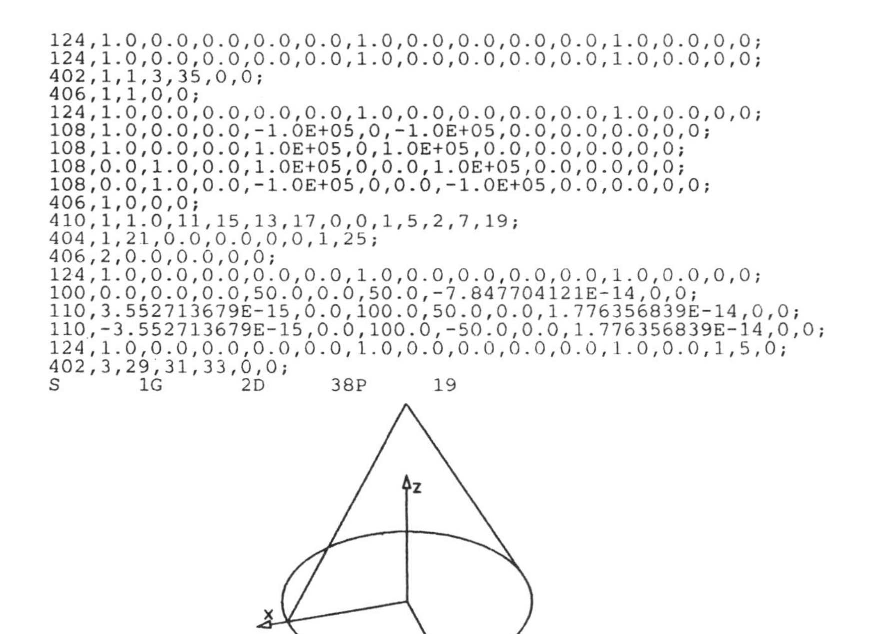

Figure 20.1 An IGES file and model

Chapter 21 Hardware data exchange standards

No data can be exchanged without some kind of hardware be it recording media or telecommunications lines. Once again, compatibility is not automatic and steps must be taken to bring it about successfully. In this chapter we consider some important hardware standards.

HALF-INCH MAGNETIC TAPE

Magnetic tape was in regular use as the principal storage medium before disks were even invented and has therefore had plenty of time to stabilise as a reliable standard for storing digital data and transferring it from one computer to another. While it is not a standard for interpreting the data in any way, it does ensure that a pattern of binary numbers stored in one computer will arrive unchanged in the receiving computer with the minimum of problems. Saying that the problems are at a minimum indicates that some exist but they are relatively easily detected and solved. The operation of a magnetic tape unit is described in "The computer configuration" on page 59.

Tape format

The format differs from that of an audio tape in a number of respects. Audio tape records the signals in four parallel tracks along the tape and uses one pair of tracks (stereo left and right) for each "side" or direction. Half-inch data tape uses nine tracks simultaneously and only writes in one direction. The stream of data sent to the tape unit is divided into nine-bit portions and each portion is written across the nine tracks - one bit per track - as the tape moves under the writing head. In some cases where the data consists of characters coded as bytes a single character is written across the tracks. The faster the portions are written as the tape moves, the more there will be per inch of tape and the larger the total amount of data there will be on the tape. There is a limit to the amount that can be packed on to an inch of tape set by the way the head is designed, the way the magnetisation is done, and the formulation and construction of the tape itself. The number of portions per inch is a significant parameter of the tape unit and is known as the *density*, expressed in bpi (bits per inch). Densities have increased very considerably over the years and the usual density is now 6250 bpi. Older units in use may still be using densities of 800 bpi and 1600 bpi. It is therefore important to ensure that two tape units involved in a data transfer can handle the same density. Many units can be switched to different densities.

The entire tape is rarely recorded in one single pass because the software will only have a certain amount of data to record at a time. When the software has data to record the tape unit will start the tape moving, write the data and then stop the unit. Such an unbroken stream of data on the tape is known as a *tape record*. The tape unit works by pulling the tape past a recording head using a rotating drum (the capstan). At the start of a record, as the capstan gets up to speed, about half an inch of tape passes before any data can be recorded. This length of blank tape is known as the *inter-record gap*. Records and inter-record gaps are shown in Figure 21.1.

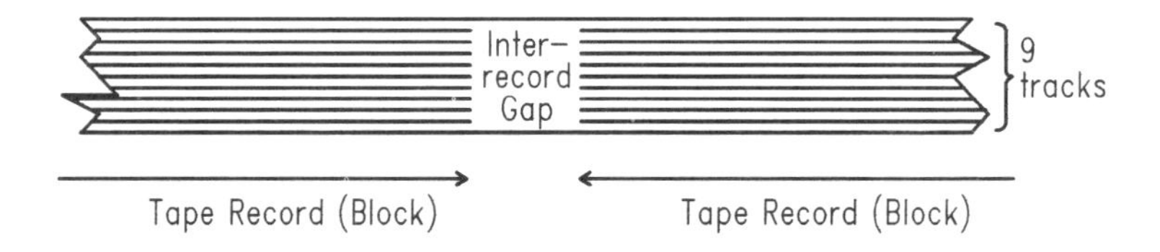

Figure 21.1 Magnetic tape format

Consider an A4 page of typescript which would consist of about 60 lines of 70 characters each or 4200 characters. Such a page would occupy 0.67 inch of tape. If a number of pages was recorded with each page in a separate tape record about half the tape would be blank. Clearly, the longer a tape record the better. Computers handle data in records which are usually in the region of 80-200 characters long, so to make efficient use of tape it would be best to write at least 200 such records in one tape record. The number of computer records in one tape record is known as the *blocking factor*. When transferring data from one computer to another by tape it is important that the receiving installation knows the blocking factor used on the tape.

Commercial data files are sometimes so huge that they overflow from one tape to another and a large data processing department will have to manage very many tapes in strictly controlled operating procedures. In these circumstances it is important for the programs to be able to write identifiers on the tape so that they can check that the right tape has been mounted etc. The practice has been standardised with formats for the identifying records laid down. Tapes written in this way are called *labelled tapes*. Unless the data transfer is being carried out within strictly controlled procedures supported by the appropriate software it is unnecessary to use labelled tapes; they can be an extra source of trouble where the data transfer process is new or under development.

Summarising, half-inch 9-track magnetic tape is a reliable and well established means of data exchange. In setting up, it is necessary to ensure that the tape densities match and that the blocking factors are known. Unlabelled tape is preferable during the initial stages.

STANDARD CHARACTER CODES

As described in "An introduction to computers" on page 5, characters are represented as 8-bit numbers or bytes. The codes used to represent them are standardised. Two standards are in common use and they give few problems. The international standard is commonly known as "ASCII" because of its origin as the American Standard for Character Information Interchange. The other has become a standard by virtue of its use by IBM and is known as EBCDIC or Extended Binary Coded Decimal Interchange Code.

The ASCII standard uses a 7-bit code, thus providing a repertoire of 128 different characters. The characters are mapped to the bit patterns of the codes in a systematic way as shown in Table 21.1 where the decimal interpretation of the codes is given. The numerals in ascending order use 48-57 so that there is a fixed difference of 48 between the arithmetical value of the numeral and its code.

Codes 1-31 are used for controlling the printer (Carriage Return, Line Feed, Form Feed etc) and for controlling transmission over long distance lines. They differ from the range used for the upper-case alphabet by 64. 64 is the decimal value of bit number 6 in the code and this has led to the presence of the Control key on the keyboard of terminals. If the Control key is held down when any alphabetic key is pressed the code generated is the upper-case alphabetic character with bit 6 dropped (i.e. 64 subtracted from the code) which allows any code from 0-31 to be generated on the keyboard. Many of these control codes have been appropriated by the writers of interactive software for special purposes, such as interrupting the execution of programs or controlling the output to the terminal etc.

Although the codes only use 7 bits, the normal unit transmitted is an 8-bit byte which leaves the coding of the eighth bit undefined. The problem most likely to be encountered in exchanging ASCII data occurs over what is done with bit 8 which can be used as a parity bit. Some programs make

Table 21.1 ASCII character codes

Decimal value of codes	Usage
0	Null character
1-31	Control codes
48-57	0 ... 9
58-64	Punctuation and typographical
65-90	A ... Z
91-96	Punctuation and typographical
97-122	a ... z
123-127	Punctuation and typographical

it zero while others make it one. The Personal Computers use it for a set of graphical symbols occupying codes 128-254. A Personal Computer reading in codes from another machine which sets bit 8 on will display text as a sequence of strange symbols.

RS232 AND V24

Purpose

The RS232 interface specification was written for the transmission of data over telephone lines. The data, in digital form, is transmitted over the lines serially, that is, one bit after another. The specification governs the connection of a computer or terminal to a data communications modem which converts the digital data into signals suitable for transmission over the lines. When used in the way it was intended it works perfectly. Unfortunately, it is also widely used to connect a peripheral to a computer and in this application, for which it was never intended, it often gives trouble.

History

The RS232 interface was originally defined by the International Telecommunications Standards Conference (CCITT) as the interface between a modem and a terminal or computer. It was laid down in Recommendation V24 which defined the functions of the signals to be exchanged. This standard was then embodied in one published by the Electronic Industries Association of the United States of America: Standard RS232-C. Since the aim was to ensure that electronic equipment could be connected together it went further and specified the electrical characteristics of the signals and the mechanical dimensions of the plugs and sockets.

An unintended application

Although originally a telecommunications standard its biggest effect was to standardise terminals and peripherals such as printers, the reason being that since any terminal could be connected to a modem it was necessary to provide it with an RS232 standard. When computers were produced and sold in quantity the natural control devices to provide were the terminals which had already been produced for telecommunications purposes and were readily available in quantity. RS232 therefore became a widely used means of directly connecting terminals and printers to minicomputers and microcomputers.

Because it was intended to cover the complexities of connecting modems to terminals in a data communications environment it is vastly over-specified in connecting terminals directly to computers and many of the signals specified are meaningless. As a result the specification is arbitrarily re-interpreted by equipment designers and its usefulness as a standard diminished.

Making it work

When used for direct connection between computers and peripherals, an RS232 interface can always be expected to give some trouble: not on electrical or mechanical grounds but in the function of the signals. This section considers some of the sources of trouble.

Ready signals

There are no less than five "ready" signals where only one or, at the most, two are needed for the simple case of controlling data flow between two devices. Because there are five signals which can be used and because their meaning is not defined for the purpose of connecting peripherals directly to a computer, equipment designers have tended to pick one of the signals arbitrarily. A device usually waits for a particular "ready" signal on a particular pin of the connector to go "on" before proceeding with the transmission. If the other device is designed to supply its "ready" signal on another pin, no transmission takes place. The problem is solved by providing an artificial signal from another pin which happens to have the right signal on it.

Flow direction

Another source of confusion is uncertainty over the direction of the signal flow. The standard covers flow in both directions simultaneously. The device on each side of the interface can both receive and transmit data but if the pin on which one device is expecting to receive data is connected to the one on which the other device is expecting to receive data, no transmission occurs. Once again the confusion occurs because the terminology of the standard relates to connections between a modem on the one side and a terminal or computer on the other side, not to a terminal on the one side and a computer on the other. To implement the specification, either computer or peripheral has to pretend to be a modem. This is done by ensuring that the data is going out and coming in on the right pins.

Transmission rates

A third source of trouble is not due to the misapplication of the standard but to the serial nature of the transmission. The bits of the data have to be transmitted one at a time down the line at a particular rate. To receive them properly the receiver needs to know the rate at which they are being sent. There are standard rates of transmission: 100, 300, 600, 1200, 2400, 4800, 9600 and higher bits/second. If the receiver is set to the wrong rate the data is received wrongly.

Data format

A fourth source of trouble is the format of the codes being sent. The data is sent in the form of human-readable characters for which there is a standard (ASCII, American Standard Code for Information Interchange). This does not give trouble. It defines a set of 7-bit codes but sometimes an eighth bit is added as a parity bit to check if errors have occurred in transmission. Some devices will therefore send eight and some seven bits. The receiver decodes the character by counting the number of bits. If it is expecting seven when the transmitter has sent eight then the data is received wrongly. Where a parity bit has been sent it can be even, odd or have no significance at all (effectively just padding). If the receiver checks the parity assuming it to be even when it is odd or of no significance then it may signal a transmission error when none has occurred.

Summary

So, to get a connection between computer and peripheral to work there are four aspects which have to be given attention:

1. Which "ready" signal does each device use, if any?

2. Which device is pretending to be the modem?

3. What is the transmission rate?

4. What is the data format: number of bits and significance of the parity bit?

The RS232 signals

Electrical characteristics

All signals are binary. The two states are:

SPACE or ON	Transmitter output between +5 and +15 volts Receiver input more positive than +3 volts Means binary 0 in the data
MARK or OFF	Transmitter output between -5 and -15 volts Receiver input more negative than -3 volts Means binary 1 in the data

Note: The data state termed MARK is the same as the control signal state termed OFF. In other words, the control signals assert the "true" condition by the ON, +ve voltage, state.

The asynchronous transmission sequence

The asynchronous mode is used when connecting computers to peripherals and in low speed data communication. The standard also covers synchronous transmission. In asynchronous mode there are periods when nothing is being transmitted and transmission can start at any time whereas synchronous transmission takes place continuously, even if just null characters are sent.

The sequence and timing of a single character in asynchronous mode is as follows. The timing between successive characters is not fixed.

1. The line is in the MARK state when not sending a character.

2. The transmission of a character starts with the transmission of one SPACE bit, known as the *start bit*.

3. The bits of the character follow, starting with the least significant bit.

4. The character is considered to have ended (by the receiver) when a known number of bits has been received.

5. The character must be followed by at least one MARK bit, known as a *stop bit*. There may be one and a half or two stop bits.

Signal functions

The main signals used between modem and terminal or computer are listed here. There are other signals which are not always used. For direct connections between computer and terminal the pins used are 2, 3, 7 and two others chosen arbitrarily.

Data

Pin 3 - Received Data	Data from modem to terminal
Pin 2 - Transmitted Data	Data from terminal to modem

Grounds

Pin 1 - Protective Ground	Earth connection provided for the purposes of electrical safety.
Pin 7 - Signal Ground	The voltage reference or return path for the signals.

Control signals to terminal or computer

Pin 22 - Ring Indicator	Indicates that someone has dialled up from outside to the modem to establish a call. The response to this must be *Data Terminal Ready* when the terminal or computer is ready.
Pin 6 - Data Set Ready	Indicates to a terminal which has initiated a call that the call has been answered.
Pin 5 - Clear To Send	Used for a one-way (half-duplex) channel where only one end can transmit at a time. Indicates to the terminal that it can use the channel to transmit after it has requested permission to transmit using the *Request To Send* signal.
Pin 8 - Data Carrier Detect	Also known as Received Line Signal Detector. Indicates that a good signal is

being received. In the case of one-way channels it also indicates that the other end has control for transmission.

Pin 21 - Signal Quality Detector Indicates if there is reason to believe that an error has occurred.

Control signals to modem

Pin 20 - Data Terminal Ready Indicates that the terminal is ready to receive the call which has just been initiated. Given in answer to *Ring Indicator*.

Pin 4 - Request To Send Requests the use of a one-way channel.

Some conventions used in practice

Full and half-duplex

Strictly, full duplex is the ability to transmit both ways simultaneously and half-duplex is the restriction of the channel to a transmission in one direction at a time.

The terms are sometimes used to describe the way that characters are echoed by the computer in interactive use. Full duplex is used to signify that each character typed is sent back again by the computer. The terminal then displays the character it receives, thus providing feedback that it has been received correctly. Half-duplex signifies that the computer does not send the character back and instead the terminal will display what has been typed on its keyboard.

Flow control

Printers cannot always keep up with the data sent to them so they need some way of temporarily halting the flow. This can be done by a selected "ready" signal but a good alternative is for the printer to send certain control characters back to the computer. There are two which have become standard for this purpose: known as "XON" and "XOFF".

Null modems

It will be remembered that the device on one side of the interface must be a modem or, if it is not, it must pretend to be one, at least to get the direction of data flow right. One way of doing this is to make up a cable in which pin 2 at one end is connected to pin 3 at the other and vice versa. Particular control signals can also be cross-connected in the same way. For example, an asserted "Data Set Ready" or "Clear To Send" can be simulated at one end using a "Data Terminal Ready" or a "Request To Send" from the other end. Such a cable is called a *null modem*.

EXERCISES

1. Why should you use a high blocking factor on magnetic tape? What would happen if you used a factor of one?

2. What character does the code 67 decimal represent in ASCII?

3. A device, A, is sending data to another device, B, on pin 3 of an RS232 interface. Which device is behaving like a modem?

Chapter 22 System administration facilities

As will be seen later, a CAD/CAM system cannot be left to look after itself. There is a number of vital operations which must be carried out regularly and with due care and attention, most of them concerned with ensuring that the electronically stored drawings generated by the users are not lost. Good drawing software is unfortunately not enough. There must be a full range of facilities to make the job of managing the system easy. Subsequent chapters will discuss how to run a system but a review of the software facilities needed for the purpose will be given here in order to complete the account of software to be expected in any good system.

DATA SECURITY

To protect against loss of work should the disk unit fail it is important to copy (*back up*) working files on to a removable medium at frequent intervals - no less than twice a day. (See also "Back-ups" on page 300.) Since there will be other files besides the working ones the file management software should make it possible to pick out the working files by a suitable directory structure.

Secondly, it is desirable to be able to make the copies without interrupting the work on the system. This can only be done if the operating system will allow two programs to run concurrently. Such a facility is one of the advantages of a multi-user operating system over a single-user operating system. The operating system must also have some kind of locking arrangement to control the situation where the back-up program and a user are both trying to access a file at the same time.

Then, to save time, it is very useful to be able to do incremental back-ups. This requires an operating system that marks a file when it has been altered and clears the mark when it has been backed up.

Regular back-ups require some kind of tape or disk rotation scheme. A utility program which keeps records and rotates the media is very useful.

ARCHIVING

Archiving is a different process to backing up although the two are often confused. They both involve copying to removable media but there the similarity ends. In archiving, the file is actually moved so that it is deleted from the working disk in order to ensure data integrity by eliminating multiple copies. Secondly, archived files are placed systematically and selectively on their removable media with the object of easy selection many months or years later.

Archiving facilities should ensure deletion of the working file once it has been correctly copied to the removable medium. They should then maintain full records of what is on each removable tape or disk, including, preferably, descriptive comment besides file name, drawing number, date last modified etc. Another useful feature is to maintain a tape rewinding or recopying procedure to avoid deterioration of data on tape over long periods.

DRAWING MANAGEMENT

A full drawing database should be available which maintains a full set of information about each drawing such as:

1. Title
2. Size
3. Drawing number
4. Modification status
5. Issued/not issued
6. Issue list
7. Issue date
8. Drawn by
9. Checked by
10. Being worked on
11. Project
12. Budget
13. Time spent on it to date
14. Plots made

Item 10 is needed because most operating systems will allow two designers to take copies of the same drawing and work on them simultaneously with disastrous results!

If a proper CIM system is in use so that drawing issue consists of passing the computer file to another department then an access control system will be needed so that issue consists of granting read rights to the receiving department.

ACCOUNTING

It may be desirable to keep records of the time spent by each designer on each drawing for budgetary control and estimating, in which case there should be suitable accounting routines built into the software. Since budgets are usually allocated against projects, it is useful to be able to record the time spent by project as well as by individual designers. As plotting consumes paper and ink or pens then it may be desirable for the accounting routines to record the amount of paper plotted as well.

CUSTOMISATION AND CONFIGURATION

Even with a turnkey system consisting of hardware and software set up and installed by the supplier, adjustments are made to tailor the software to the particular needs and practices of the individual customer. Furthermore, the supplier usually offers a number of alternatives in hardware and the customer will wish to upgrade with additions from time to time. For every hardware alteration there is usually an adjustment to the software by way of additional driver routines or parameters in the driver routines. All this requires utility programs to make the necessary alterations.

A typical system has a very large number of customising parameters varying in the extent of their effects. Firstly, there are the settings which apply to the individual workstation. If a tablet menu is in use, there will be parameters recording the position of the menu on the tablet, options on the size of tablet or whether a puck or pen is used. The screen may need describing: colour or mono and resolution.

Then there are the settings which apply to each user or department: special menus, working directories on the disk and parameterised drawing programs.

Finally, there are the settings which apply to the installation as a whole: plotter driver, software options, fonts, line types, hatch patterns, dimensioning standards and the structure of the drawing management system.

New settings should always be documented, preferably by the utility program itself either by printing the settings out or recording them in a text file. The effect of some customising parameters, particularly in relation to the graphics display, can be quite obscure. The parameter is set up on installation and is then forgotten about until an inadvertent error by the manager alters it and the software starts to behave oddly for no apparent reason. The management of such a large number of different software parameters requires careful records to be kept and the more assistance given by the software the better.

PLOTTER DRIVERS

All CAD/CAM vendors like to offer a choice of plotters. In order to do this the vendor can provide a plotter driver program for each plotter. Alternatively, he can provide a general-purpose driver which has everything in it except the actual output subroutines which pass data out to the plotter. In this case the plotter manufacturer is expected to provide the output subroutines and, as part of the installation process, they are compiled or link-edited with the CAD vendor's program. The functions and calling format of these output subroutines have become standardised within the industry and conform to one introduced by the Calcomp Corporation.

EXERCISE

Summarise, in your order of priority, the facilities the software should provide to make the management of a CAD/CAM system efficient.

Chapter 23 Robustness, reliability and support

In an engineering department genuinely committed to CAD, failure of the system will not just be a nuisance but it can cause the work of the department to come to a halt. As a consequence, any feature or facility which reduces the chance of the installation failing, or even shortens the period for which it is out of action should it fail, is very important.

It is obvious that the system should be reliable, meaning that it should not fail under normal use but there are other characteristics which have a bearing on the matter. The operators, being human, will treat it incorrectly from time to time by inputting wrong commands or data or by allowing the hardware to experience extreme physical conditions. It should be able to withstand a certain amount of such ill-treatment and we can call this characteristic its robustness.

Besides minimising the occurrence of failure the system manager needs to consider what happens when it occurs since, like it or not, there will be failure on occasions and there are characteristics which affect the course of events after a failure has occurred. A simple but important one is the time taken to rectify the fault or the down-time. Now the failure will occur in just one component of the system. What will be the effect on the system as a whole? Will it stop the system completely or will it simply make the operation less efficient and slower? This is another kind of robustness: the ability of a system to withstand failure in any of its components. A mode of failure which only reduces performance is known as "graceful degradation". Sometimes it is possible to configure a system in such a way that this is the most likely effect of any hardware failure. The usual way is to duplicate items of equipment. For example, a system composed of two computers working together has the same functionality as a system composed of one but with twice the speed. However, if one fails the system will simply slow down but not come to a halt, as it would with one high performance computer twice as fast. The advantage of a system of individual workstations, each with its own computer, in this respect is clear although even in such a case there is usually a single shared file server on which the data is stored. Finally, another vital aspect of robustness is that failure should not result in data being lost.

The main aim of the system manager is to provide the maximum performance for the maximum time or to maximise what can be called the availability of the installation. Reliability, robustness, down-time and the severity of the effect of failure all combine to affect this objective of maximum availability.

HARDWARE ROBUSTNESS AND RELIABILITY

Computers, being electronic devices with very few moving parts, are very reliable. As you might expect the main areas of failure are where there are moving parts. They are usually cooled with filtered air blown by fans. The filters can block up unless cleaned or replaced. Tape drives have a lot of moving parts and require precise mechanical adjustment which can go off. The most vulnerable part of a computer is its disk drive. As described in "The computer configuration" on page 59, disk drives require carefully filtered air and can fail in such a way that all the data is lost. Sealed disks ("hard" disks or "Winchester" disks) will be more reliable because dust particles cannot enter. Note that the very component on which your drawings are stored is the least reliable of all, hence the need to back up data on to removable media. The software for doing this makes an important contribution to the robustness of the system.

The other hardware cause of data loss is power failure. Some or all of a drawing currently being worked on is held in the volatile fast semiconductor main memory of the computer. This data disappears almost instantaneously when the power fails. Along with this data are internal variables used by the disk file management software to keep track of the disk file currently being accessed. Fortunately, the design of the power units in the computer prevents the power disappearing before it can perform some emergency routines to tidy up a little first. It is possible to design the operating system and the CAD software so that the data lost on power failure is negligible. Unfortunately, this is not always done and is a point to check when evaluating systems. The alternative precaution is to install uninterruptible power supplies which keep the computer going on batteries sufficiently long for data to be stored safely on disk.

An important aspect of hardware robustness is error detection so that any malfunction in the computer circuits is detected before it can cause data to be corrupted. Corrupted data is more serious than lost data since it cannot be detected easily. Diagnostic routines are usually built into the computer and executed every time it is powered up. In some systems they are executed automatically in idle moments.

HARDWARE SUPPORT

The down-time is directly affected by the speed with which a defective component can be replaced or repaired and this in turn depends on how soon a service engineer arrives to do the work, whether he brings the spare part with him or how long he takes to obtain it, which in turn depends on the size of his company's stock of spares. All these vary from one company to another and from one part of the country to another.

There are various ways of procuring support. How and where to obtain it and how much to pay for it needs careful consideration. The usual way is to sign a contract with the original supplier: for an annual fee he will perform preventative maintenance (cleaning air filters etc) and guarantee to have an engineer at your site within a specified minimum time of receiving a request and to replace defective parts without further charge. The contract has a lot in common with an insurance policy, the fee being a premium covering the cost of failure, so that the size of the premium is related to the amount of risk borne by the supplier plus a charge to cover the cost of holding a stock of spares.

The outlay on maintenance can be reduced in a number of ways. The biggest saving is achieved by not having a contract at all. You could do the preventative maintenance yourself although you may be at risk in not being able to diagnose faults or find spares. However, faults in some equipment can be easy to diagnose, particularly with good diagnostic software, and there may be circumstances where it pays to hold spares. If you are not prepared to do diagnosis then you can simply call in a service engineer from the supplier or a maintenance company when the fault arises and pay for his time and the spares he has to supply (or draw on your own stock of spares). The cost will seem high but may still be less than a maintenance contract taking into account the frequency at which faults occur. Because there is no contract between you and the maintenance company there can be no guarantee to fix the fault by any particular time, although the maintenance company may wish to uphold a good reputation for the sake of further business.

If you have decided to have a maintenance contract you may be offered options in it which allow you to save money. Some companies offer a range of grades of service representing different degrees of response time from, say, arrival within 3 hours to arrival within 24 hours. Another variant may be whether the contract covers the cost of spares or whether you pay for the spares at the time of breakdown.

Because of the opportunities for reducing running costs the maintenance arrangements for each part of the system should be decided separately. Obviously, items which will halt the whole operation when they break down need to be on a maintenance contract with a carefully chosen grade of service, but there will be items which do not fall into this category for which less expensive arrangements can be made. Some things can be left without any maintenance arrangement at all because they are either inherently reliable or you can manage without them for a while. Duplicated items present particular opportunities for saving as it is often possible to continue at reduced performance with one item out of action while a service engineer turns up or even while you diagnose the fault yourself and procure a replacement. Fault diagnosis by substitution can be performed on duplicated equipment. Suspected components from the faulty item are interchanged with identical components from healthy items to discover the faulty component. Even more useful are redundant items of equipment held as spares to bring in when there is a failure.

SOFTWARE RELIABILITY

It is a strange paradox that software which has no moving parts to wear out breaks down in a random fashion just like machinery. The reason was touched on in "An introduction to computers" on page 5. Any large complex software like a CAD/CAM program has an enormous number of different paths through it by virtue of its ability to take decisions. The number is so large that it is impossible for the programmer to test all possible routes and so the program is issued to users with one or two defective routes containing errors. During normal use, different routes are taken at different times as a result of the many different commands and data values input by the users. Some routes will be taken more frequently than others so that each route has a particular statistical chance of being taken. Errors in frequent routes will be discovered during testing but there will be routes which are only taken very rarely. The first time one of these is taken will be a chance occurrence and if it contains an error the software will appear to fail randomly and unexpectedly, since the user is unaware of the many decisions the program is taking inside itself.

So we have the situation that any complex program will have errors or "bugs" lurking in it which have not been discovered during testing and which will manifest themselves gradually in a random fashion over a period of use. How can the incidence of bugs be reduced? Obviously, by taking care in writing the programs in the first place. Much work has been done and is still under way on devising programming languages and programming practices which reduce the chance of error. On the human side, small teams seem to produce better quality software than large ones. Having paid attention to the quality of the program and tested it thoroughly before issue, further steps can be taken to improve the quality after issue. On account of the random nature of the occurrence of bugs, the more the program is used in as varied a way as possible, the more bugs will be found. A large number of regular users is clearly important. A bug manifests itself by certain symptoms, that is, unexpected consequences from a particular action on the part of the user. When they occur, these symptoms and the action which induced them must be recorded and passed back to the programmer who can then locate the error in program code and correct it. After a number of bugs have been handled in this way the supplier issues a corrected version to the users.

Summarising, the conditions for good quality software are:

1. Good programming practice and technique

2. A large number of regular users

3. A bug-reporting procedure from user to programmer

4. Regular bug-correcting issues

With these in place the quality of the software should gradually improve. Two factors hinder the improvement. Software is currently sold on its features and not on its quality. Software suppliers must therefore be continually producing new code with new features to stay competitive. The new code will introduce new bugs. Also, the supplier has to decide how to allocate his programming staff between bug correction, which only keeps existing customers happy, and new code, which keeps his business expanding. The other factor which hinders the process of quality improvement should be rarer as programming practice improves. Unless programs are designed properly it is possible for a situation to arise in which each alteration to the program done to correct an error introduces another error somewhere else, rather like a "do-it-yourself-er" who refixes a shelf which has fallen off the wall by drilling another hole which happens to go through a water pipe buried in the plaster!

The quality of software can be measured in terms of the number of bugs per thousand lines of source code. For example, one very large CAD/CAM package was quoted as having 1.5 bugs per thousand lines. The incidence of failures experienced by its users was enough to reduce efficiency and cause comment. When such a figure is quoted it should not include a count of the comment lines in the source as these, of course, cannot cause bugs!

It was stated earlier that a bug appears as an unexpected consequence of something the user does. The user's expectation of what will happen in any particular operation is set by the training he has received and his interpretation of the manuals. This means that even if the program is doing what the software team intended it to do there is still a "bug" if it does not do what the manual says it will do. (The term "undocumented feature" has been used on occasions as a euphemism for a bug.) So, bugs can be caused simply by badly written manuals or bad training. Training is covered in more detail in "Training, manuals and user groups" on page 287.

SOFTWARE SUPPORT

As can be seen from the foregoing discussion, bug correction, documentation and training are very closely bound up together. When the system does not behave as the user expects it can be because:

- The program has an error
- The manual has an error
- His training is incomplete

Since correcting programs and manuals or training users costs money, it has become common to provide a software maintenance contract for an annual fee rather like the hardware maintenance contract. A good contract on a package which is under steady development will provide new issues once or twice a year to correct bugs, new issues of manuals, a telephone permanently manned to answer problems and to receive reports of bugs, and completely new issues of the package with new features every two years or so. The most important part of any support service is the telephone manned to answer queries, since many of the problems will be due to inadequate understanding of how to operate the software and can be dealt with immediately. No CAD/CAM software should be bought from a supplier who does not have such a "hot line".

Support is provided a little differently in the case of Personal Computer packages sold in large numbers through a dealer network. The supplier will be able to carry out thorough user testing before finally releasing the package on the market. The level of bugs is then sufficiently low to be of little trouble to the average user. The large market will mean that even these will be quickly found and corrected versions distributed through the dealer network. The scale of the operation allows more to be spent on producing the documentation and this will be supplemented by support from the dealer, telephone support, independent training companies and the technical press. New issues with new features are offered as separate products with a special upgrading price to those in possession of the previous version.

SOFTWARE ROBUSTNESS

There is a big difference between an amateur programmer and a professional programmer. The amateur will be the only person using his program and he knows its limitations well enough not to input silly data. The professional has no control over what will be fed into his program and has to make sure that it behaves properly no matter what rubbish is fed into it. A large part of any professional program is devoted to checking input data and deciding what to do with values which are undesirable, either because they will cause the subsequent calculations to break down on, say, a division by zero or because they cannot possibly represent the value being asked for and would produce a misleading result. A part of robustness in software is this capacity to deal successfully with bad data. Bad data should be recognised by the program and then handled so that:

- It does not corrupt the data already fed in
- The user knows what his mistake has been
- The user is able to correct his mistake easily

The last requirement is not always met. Sometimes the error occurs in a long sequence of values. It is a very bad dialogue design which requires the user to re-enter the whole sequence again just to correct the erroneous value.

The other part of software robustness is the ability to limit the effects of malfunctioning hardware and this has been discussed earlier in the chapter.

EXERCISE

Summarise, in your order of priority, the factors needed to make a system reliable.

Chapter 24 Making the case

The benefits of CAD/CAM were discussed in "CAD/CAM and its value" on page 13. Here we will look at the topic from the angle of making a decision whether or not to procure a system. Besides a consideration of the benefits, making the decision involves the political dimension - persuading all concerned to invest a substantial sum of money and to go through the upheaval resulting from a major change in technique. Lindgren in Reference (36) draws attention to these very real political factors which are often ignored by technically trained people. We will illustrate the political aspects with some imaginary but not too far-fetched stories. We start with the ideal sequence of events for comparison.

THE IDEAL SCENARIO

This is an ideal sequence of events which many companies will be able to follow. The initiative comes from the Managing Director who, realising the growing use of CAD/CAM, decides to carry out an investigation. He sets up a small team composed of someone from Production and someone from Engineering. They visit other companies using CAD/CAM, and read books and case studies in journals. They produce a report which:

1. Indicates which benefits are applicable to the company.
2. Estimates the value of a benefit where it is quantifiable.
3. Indicates which activities in the company should use CAD/CAM techniques.
4. Estimates the cost, duration and manpower requirements of a procurement exercise.
5. Estimates the likely capital cost of the equipment.
6. Proposes the composition of the team for conducting the procurement.
7. Supplies outline programmes for procurement, installation and training.

8. Estimates the duration and severity of the loss of efficiency during installation and training.

With this information the Board approves a budget for conducting the procurement up to the point at which equipment is recommended for purchase. The procurement team gets to work and produces a recommendation which the Board approves. With support by the whole of management from the top and a properly considered programme for installation and training, the company is well placed to go through the severe disturbance of a major change in its operating methods. The installation proceeds. The commitment of all concerned means that the use of the equipment is not half-hearted and every effort is made to obtain the benefits the original study indicated. The installation plan includes a trial period at the end of which the actual achievements are reviewed. After the review, any application of CAD which is not fulfilling its promise is abandoned.

THE DILETTANTE SCENARIO

The company has made a large profit as a result of some very good contracts. It is looking for ways of investing in equipment. The Managing Director, noticing a number of firms procuring CAD equipment, decides to do so as well, his justification being that the company will get left behind if it does not. He calls in three CAD suppliers recommended by his friends at the golf club. Each supplier makes a presentation to selected members of staff and provides a quotation. A meeting of managers decides to recommend the Board to purchase a particular make - the one which happens to have been recently chosen by a friend of the Technical Director. The equipment is duly installed in a room next to the Design Office, there being no space available actually in the Design Office. Two recent young recruits to the Design Department are sent on the training course and then use the equipment in their spare moments along with people from the Research Lab and the Technical Director himself. During the first six months its principal output are plans for a new extension to the latter's house and all the posters for the Social Club events. It is never used by the engineering designers. They are all too busy on urgent work to spare the time. After three years the company's fortunes have taken a downward turn and it is bought out. An accountant in the new parent company comes across the annual maintenance bill for the installation and asks what the company is getting in return. The equipment is sold off.

THE POLITICAL SCENARIO

The Chief Designer in a largish company keeps himself well informed of all the latest developments by means of journals, the occasional book and local meetings of his professional institution. It becomes apparent that CAD, more than just a fascinating new technique, could actually benefit his company. With the support of his Engineering Director and a considerable amount of research, he prepares a case for purchasing the CAD equipment of supplier X. This shows that it would pay for itself through cost savings in no more than three years and would result in better and more timely proposals in bidding for new work. The case is submitted to the management and from informal talks with senior managers he is pleased to learn that it has been received with great interest but nothing further happens. After many months he has to conclude that he has failed to get any action. He pursues the matter no further and thinks bad thoughts about the way engineers are despised and about the evil influence of "politics".

What went wrong? He was correct about the influence of politics but it was hardly evil. The trouble was that he was proposing the purchase of additional computer equipment. The management naturally referred the proposal to the accepted computer expert, the Data Processing Manager. He read it but a number of things made him unenthusiastic. The company's computer equipment came from supplier Y. He had always used supplier Y, had got his present job because of his long experience of supplier Y and he did not know very much of supplier X. Actually, the sales representative for Y had once suggested he bought his company's CAD package but it was prohibitively expensive. The cheapness of the Chief Designer's proposal together with the unknown status of supplier X made the proposal look risky and amateurish. What did the Chief Designer know about choosing reliable, professional computer equipment?

Besides this, there was a long list of enhancements the Data Processing Manager himself had been trying to get approved. He was under pressure from the Production Manager on account of the slow response times of the stock control system and with his support he had put up a case for additional hardware to improve the situation. The cost saving resulting from the additional productivity of the data entry staff produced by the faster response had been calculated and it paid for the cost of the hardware over three years. The cost saving offered by the Chief Designer's proposal was based only on estimated figures for faster drawing times. In any case, there was no data on the current productivity of the designers for comparison. As regards the argument for better proposals that could not be related to cost benefits.

The Chief Designer's case failed because he was unaware of political factors and even regarded them as undesirable, or at least as a necessary evil. But arising as they do out of human nature and human relationships they are more fundamental than technical issues.

We will now re-run the scenario with a politically aware Chief Designer to illustrate the political devices available to those wishing to make a case. Before writing his proposal he thinks who is likely to have the most influence over the outcome and realises that the Data Processing Manager is likely to be very influential. He arranges to meet the Data Processing Manager and puts over his ideas for a CAD system as persuasively as possible. Advocacy is a fundamental political procedure. The Data Processing Manager's enthusiasm is caught but then he cools off. He talks of the problems of getting funding and of all the enhancements he already has on the table and mentions his misgivings about the Chief Designer's supplier X. The Chief Designer offers to consider supplier Y's CAD offering instead. Negotiation, in which one gives way in one direction in order to gain in another, is a frequent political procedure. As the Data Processing Manager talks about his own proposed enhancements he mentions the stock control data entry problem. The Chief Designer spots an opportunity. He points out that the CAD system can pass parts lists directly into the stock control system, thus reducing the data entry load considerably. He suggests the two of them getting together with the Production Manager to put in a joint proposal. Building coalitions is another useful political device. The Chief Designer has allied his interests to those of the influential Data Processing Manager and the powerful Production Manager.

The Chief Designer's discussion with the Data Processing Manager has made him realise something else of importance. It has become the accepted practice in the company to justify computer enhancements by cost savings, which is in fact a weak part of his case since there is no well established data in design about productivity. The stronger part of his proposal regarding sales proposals is not based on any kind of cost saving. He therefore arranges to see the Sales Manager. It does not take long to get him enthusiastic over the prospect of better and faster bids. The Chief Designer now has strong support for his proposal from three influential people in the company and is well on the way to success.

THE VARIOUS KINDS OF BENEFIT

In making a case, various kinds of benefit are arrayed in support. There is often muddled thinking and confused terminology in stating benefits. An astute management will be aware of the different kinds and know which ones count. Financial benefits are often called "cost savings". This term is rather vague and a number of more precisely descriptive terms will be considered.

Cost displacement

Cost displacement is the reduction in some recurring operational expenditure. More specifically, it usually means a decrease in a budgeted item. In other words, the introduction of the computer facility will result in the particular budget being reduced. The item being reduced may be headcount, supplies, bank charges or inventory carrying costs etc. Because it is quite easy to evaluate and monitor, cost displacement is widely used as a benefit.

Improved efficiency or productivity gain

Related to cost displacement is the argument that the computer facility will get the job done with, say, 20% less labour. A financial saving is then calculated by multiplying the 20% by the labour rate. It should be remembered that the benefit only occurs if the time saved is actually spent on other useful work. Productivity gains have been bandied about extensively in recent years with reference to CAD systems but are beginning to fall out of favour for very good reasons. The figures have been used in a way which gives an excessively optimistic picture. To start with, instead of quoting a percentage labour saving - a fractional quantity - the gain has been quoted as a ratio of the time required to do a job without the computer to the time required using the computer - usually a number between 1 and 10 - which sounds impressive. If R is this ratio then the labour required to do the job originally was R units while with the computer it is 1 unit. The saving is $R - 1$ units which as a fraction of the original labour is:

$$\frac{(R-1)}{R} = 1 - \frac{1}{R}$$

expressed as a percentage this is:

$$100 - \frac{100}{R}\%$$

Mechanical drawing usually has a ratio of 1.5, which gives a labour saving of 33%. The figures are further misleading if those reading them are used to the cost displacement of full computerisation in which the entire job is done by the computer. With CAD the job is done by someone interacting with the computer. For a job which takes 1 unit of time using a workstation and R units of time by a man on his own, if M is the cost per unit time of a man and W that of the workstation then the cost of the job with the workstation is the cost of both man and workstation: $M + W$. The cost of the job done by one man on his own is R times more: RM. For there to be cost displacement:

$$M + W < RM$$
$$\therefore\ W < M(R - 1)$$

It is not sufficient for the workstation to be less than the cost of a man as it would be in conventional computerisation. For example, taking the mechanical drawing example above where $R = 1.5$, the workstation must cost less than half that of a man before there can be true cost displacement.

Further misconceptions can arise if the management thinks that a designer spends all his time drawing and imagines that the workstation is in use all the time. In fact, the workstation cannot be fully utilised. Even if shared between several designers so that when one is not drawing another is using it, the utilisation will be less than 100% due to inevitable inefficiencies in scheduling its usage. The utilisation directly affects the cost per unit time, W, of the workstation since W is calculated by dividing the annual cost by the time per annum for which it is profitably used. For this reason, some companies like to run a simple shift system to increase the time for which the workstations are in use.

Increased revenue

As indicated, one has to work hard to show cost displacement through improved productivity but this is not the case when increased revenue is the objective. A small competitive edge can tip the balance and bring in more business and, for the reasons given in "CAD/CAM and its value" on page 13, CAD has the potential for tipping the competitive balance. It therefore differs significantly from the traditional computer applications which only affect the internal operations of a company. For this reason the justification methods used for data processing facilities should not play too big a part in the case for CAD/CAM.

Cost avoidance and risk reduction

In budgeting for running costs there are expenditures, such as repairing machinery, which have to be allowed for even though they may not occur. Removing the necessity to allow for them or reducing the chance of their occurrence is a benefit sometimes referred to as cost avoidance, although a better term would be risk reduction. An example of cost avoidance in CAD is the reduction of errors in drawings as a result of the use of automatic dimensioning calculated from the accurate numerical model which constitutes a CAD drawing.

Although undeniably a benefit, risk reduction does not remove any actual known cost and is therefore very difficult to quantify. For example, to quantify the effect of automatic dimensioning there would need to be a record of the incidence of errors in the past and the cost of rectifying them. Then the incidence of errors after introducing CAD would have to be monitored and a comparison made.

Unquantifiable benefits

Although it is usual to think of all benefits as ultimately resulting in reduced costs or increased revenue, there will be benefits which do not have an obvious or immediate connection with cost or revenue. They should not be ignored on that account but presented in detail under the heading of "unquantifiable benefits". Those reading the proposal can then judge the financial effect for themselves. Sometimes the fact that the benefit will help them do their job better will be sufficient. Examples of such unquantifiable benefits are design analysis facilities, high quality drawings for presentations and proposals, perspective projections and colour-shaded pictures of surface or solid models.

Improved service

One of the more important objectives or consequences of unquantifiable benefits is improved service to the firm's customers. This can either take the form of offering more without the firm incurring additional cost or removing existing sources of irritation which were too expensive to remove before. Faster responses to requests, better quality drawings and more accurate and fuller data are possibilities in this area.

Improved image

Because of the high technology involved and the spectacular nature of CAD, some companies have installed it largely to project a picture of a progressive company. If the system does no more than pay for itself the company has got the better image for nothing. This can be done by locating the workstations where visiting prospective customers can see them or by doing proposals in a way which shows that CAD has been used. The value of using CAD in this way will of course disappear as it becomes commonplace.

Transfer costs

In large organisations it is common to have one department charge another for goods and services such as secretarial services from a typing pool, or production materials supplied to one plant by another. The transfer costs may seem like real ones but their rates have been set artificially. At one extreme, the price may have been set at what would be charged by an external source in order to force the unit to be competitive with outside sources. At the other, the price may simply be the variable cost of providing the item or service. Saving these costs does not save the company as a whole any money and if money is spent externally to save them, such as on a CAD system, any benefit is an illusion.

THE RUNNING COST IN COST-BENEFIT CALCULATIONS

An important part of cost-benefit calculations is the running cost of the equipment. This is composed of the maintenance cost plus the cost of ownership. The maintenance cost is fairly easy to obtain, being usually about one-tenth of the purchase list price per annum, for a maintenance contract.

The cost of ownership, on the other hand, depends on how the equipment has been paid for and how one depreciates it. Depreciation assumes that the equipment is wearing out continuously and will have to be replaced by equivalent items at equivalent prices. Computer equipment, however, has very little in it to wear out. Furthermore, the cost of equipment of equivalent performance is dropping all the time. Should one depreciate equipment which will always provide the service for which it was purchased and for which the cost of replacement is always dropping? The main thing which is changing is the expectation of what such equipment should do. Although it will always provide the service for which it was bought, it will rapidly appear obsolescent and its resale value will drop. A judgement has to be made on depreciation before one can do a cost-benefit analysis.

CONCLUSION

In this chapter we have shown the good and bad ways of handling the political aspect of making a case, analysed the various kinds of benefit which could be used to support the case and considered the factors involved in the running cost. We have shown that it is important not to ignore the political dimension, to get support at the top if possible and to realise that

CAD/CAM gives benefits which are of a different kind from those of traditional data processing.

EXERCISE

In the dilettante scenario suggest some measures that could have been taken to save the situation.

Chapter 25 Management change

It is a mistake to think that the installation of a CAD/CAM system is just like putting new drawing boards into the Design Department. It does not simply make the department a bit more efficient. It radically alters the way it works, affects the way other departments work with it and offers opportunities for improvement throughout the company. In this chapter we consider the impact of CAD/CAM on the company as a whole. For a fuller treatment of management issues throughout the company see References (28) and (29).

TECHNOLOGY CHANGING MANAGEMENT

CAD/CAM epitomises a process of change which started when a computer was first used to do accounts and which has been gathering momentum ever since. The use of electronic devices has grown steadily over several decades until we have reached the point where they have become essential tools for most "white-collar" workers. The process can be compared with the development of steam and then electric power on the shop floor. Just as steam changed the factory floor so computers are changing the office. In both cases the change is structural: not superficial. Information technology is changing business management at the day-to-day level. To manage this change there must be an understanding equally of management and of the nature of the technology which is forcing the change. It is not good enough to leave the technology to the specialists who by their nature cannot see the position from a management point of view. Every company needs a new breed of managers who understand the technology although not to the same depth as the specialists. Obviously, one cannot exchange one's managers for a new set overnight but at the very least every company should ensure that there are a few people who have this dual understanding either through recruitment or training. The rest of the chapter underlines in some detail this basic need for a technologically educated management which will square up to the changes and manage them.

CHANGE WITHIN THE DESIGN DEPARTMENT

The designers

Most of the quite radical changes that have occurred in the world during the last three centuries can be attributed to engineering. It is therefore quite appropriate that engineering designers should have to experience the most dramatic technological change of all in the tools of their trade. After years in which the only advancement has been the invention of the clutch pencil they are now having to give up their simple boards and sheets of paper for some of the most complex computer software and hardware known to man! The revolution in their tools is so great that even their working methods have to change.

Any company which is serious about introducing CAD/CAM has to face up squarely to the consequences of the revolution it will inflict on its Design Department. For each designer there is going to be a period of several days while he trains when he will be capable of no useful work at all. Following that, there will be a significant period during which he will be working at less than his normal efficiency while he becomes accustomed to the unfamiliar equipment. Some of the designers may never become accustomed at all and may have to opt out of using it altogether. All this will take place in a department which is playing an important role in the work of the company and may well be under some pressure.

To prepare for such a change, so that the bad effects are minimised, requires the full commitment of the senior management of the company. It is too serious a matter for Engineering to sort out on its own. The upheaval is going to cost money in terms of lost time and the Design Department is going to have a reduced capacity during that period. The effect can be minimised by proper phasing of the operation and the provision of additional labour, if necessary by extra working hours or subcontracting. Financial allowances will have to be made.

The supervisors

So far we have been considering the impact of CAD/CAM on engineering designers but it will have just as great an effect on those who supervise design in a number of different ways. Wells describes the changing position of the design supervisor in References (3) and (31).

Traditionally, the supervisor in most walks of life is one who commands authority because he has more experience than those he supervises. When a CAD/CAM system is installed, everybody who has been trained to use it is equal in their experience so that in one respect the supervisor has been levelled down. His management may even decide not to train him because he is too busy to be spared.

One of the uses to which the experience of the supervisor is put is in estimating the work or time scale required for a job. The CAD/CAM system makes any previous standards for estimating jobs obsolete and experience in estimating has to be developed all over again. In addition, new standards for assessing the performance of individual designers will have to be developed. The problem will be exacerbated if the system has been justified on a productivity basis since the spotlight will be focussed on the time taken to do jobs.

Previously, the task of allocating work only needed consideration of the skills of the various designers and their current jobs. With a CAD/CAM system running in parallel with traditional methods the suitability of the job for CAD/CAM and the availability of workstations also has to be considered.

A new way of monitoring progress will have to be used. It is no longer possible for the supervisor to get an idea of progress by walking down the row of boards looking at what is on them. The screens will be too small to see properly and in any case the whole drawing will not be visible on them. Also, discussion on a design cannot be carried out in the same way on a screen as it used to be done by marking up a print. An alternative way of monitoring work is for the supervisor to bring copies of drawings that designers are working on up on to his own screen. He could mark up a copy on his screen for the designer to look at on his screen. It may still be best to take plots and mark them up in discussion as before.

Sign off and issue will be different also. Less checking should be required before issue because of the precise numerical model on which a CAD drawing is based. CAD drawings cannot take a signature. The computer equivalent to issue is the altering of access rights to the file. The computer user ID used by the supervisor would be given the privilege of altering access rights. The issue operation would grant read rights to the departments receiving the issue and deny write access to all since, once issued, a drawing cannot be altered without raising a new drawing number or at least a new modification status to the drawing, which is effectively a new drawing.

Although most CAD/CAM software is written to minimise the amount of specialist computer knowledge required, there are a number of computer-related activities needed to run a system such as:

- Data security and storage management
- Installation of new software issues
- Parameterised drawing programs

- Selection and stocking of consumables
- Reporting and fixing of bugs
- Training

They involve a deeper knowledge of computers beyond that needed in the daily use of the system for design, so rather than train all the designers for these special activities it is natural to train or recruit just one man and give him responsibility for them. This leads to the position of "CAD/CAM Manager". The position is a new one for a Design Department in a very important respect. An additional person is now introduced who has specialist knowledge that the supervisor does not have but who is important to all the work of the department. The tool used by the designers has become a speciality in itself.

The implication for management is that the relationship between the roles of design supervision and CAD/CAM system management will need to be thought through and properly defined, otherwise there will be conflict: either overt or hidden. If his management has not considered the matter already, someone appointed to the position of CAD/CAM Manager will need to think about his relationship with design supervision. Potential areas of overlap are the selection of work for the CAD/CAM system, training and the new drawing office standards made necessary by the system.

So far, our discussion of the position of design supervision has been negative. There is a positive side. If the design supervisor is included in the procurement team he will be an important source of information about existing procedures and practices, and will be able to provide useful guidance on any new procedures needed. In the procurement team he will have an opportunity to learn about the new technology and develop a commitment to it, which will be passed on to the designers. During implementation he will play a key part in phasing in the new system and if trained himself will be a good person to do training. If senior management is looking for greater efficiency the design supervisor is the best person to achieve it because of his experience with motivational problems.

CHANGE OUTSIDE THE DESIGN DEPARTMENT

The Design Department does not exist on its own and for its own benefit (whatever some cynics may say!). It provides designs for Production to manufacture, specifications on which Buying have to place orders, data for the stock control system, the designs which Marketing sell and Senior Management have to appreciate, data for Costing and information for

Field Servicing, Customer Support, Customer Training and Technical Publications. Building Services will be providing special cooling equipment, fire detection equipment and cabling for the CAD installation. Designers, being employees, are the concern of Personnel and Finance will be very much aware of the substantial capital investment made in a department which hitherto had only labour costs associated with it and which is not directly connected with the actual money-earning parts of the company.

There is no doubt that the arrival of CAD/CAM changes the company. We will now look in more detail at the changes CAD/CAM could bring about in the various parts of the company.

Production

CAD/CAM provides the opportunity to have more accurate information because it has been obtained directly from the CAD drawing file without human intervention. If Production has computer systems then there is the further opportunity of reading the data in without rekeying by a human. The dimensions on the drawing will have been generated automatically by the CAD software from the geometry of the drawing, thus reducing the opportunity for human error escaping the checking process. If NC machining is in use the geometry of the part programs can be extracted from the geometry of the CAD drawing leaving the part programmers to concern themselves with the actual machining process. Parts lists can be read out from the drawing without human intervention, thus reducing the opportunity for error.

Assembly instructions are often helped by perspective drawings which were too time consuming by manual methods but which are rapidly produced in a CAD system.

Costing will be assisted by accurate and timely parts lists and in certain cases by calculations of volume, weight or surface area performed by the CAD system.

Considerable research work is currently being undertaken into ways of helping designers to reduce manufacturing costs in their designs. Some of the methods being investigated consider the use of software in the CAD system which is programmed with rules for reducing manufacturing cost and which can examine the CAD drawing directly and suggest improvements. References (32) and (33) describe some of the current work in this area.

It can be seen that the effect of CAD/CAM on the Production Department is largely to provide opportunities for increasing efficiency rather than to force anything on them. The department must investigate and take up those opportunities which are relevant in its particular circumstances.

Building services

The involvement of the Site or Building Services (or Plant) Department in providing special services to the CAD/CAM installation has already been mentioned but CAD is a very useful tool in laying out new plant. It allows different arrangements to be tried out rapidly and the selected arrangement to be drawn out rapidly with the minimum of labour. The use of layers allows all services to be drawn on the same drawing and viewed selectively. Data extraction methods can calculate loads on services, floor loadings, clearances etc.

Buying

Parts lists and specifications can be produced in computer files by the Design Department. This offers the opportunity of producing purchase orders directly from these files or reading specification data directly from the file produced by the Design Department into the word-processor preparing the tender or contract documents.

The biggest impact on the Purchasing Department will occur when CAD drawings on diskette are sent to subcontractors instead of paper drawings (as is already being done by some companies). The contractual implications in relation to proving that the data sent was correct and in ensuring that it is treated with proper confidentiality need thinking through carefully.

Sales

There are some companies where the biggest benefit of CAD is obtained in the Sales Department rather than in Engineering. CAD produces high quality drawings for proposals very rapidly by modifying previous ones or by the powerful parameterised drawing technique. Solid modelling and powerful colour-shaded picture generators produce realistic illustrations which give confidence in the company's ability to deliver. Data extraction programs linked to symbol libraries make it possible to design a new layout and cost it up automatically. This is particularly applicable to the capital goods industry bidding to carry out new plant installation and similar business. Some companies have even bought a CAD system principally to create an up-to-date image!

Outside the capital goods business, Marketing will appreciate the faster time to get a new product launched that is possible with CAD/CAM and better designs within the shorter time scale. If the opportunities of design for manufacture provided by CAD are taken up then reduced manufacturing costs become possible.

Customer services

Maintenance drawings or illustrations in operation manuals all have to be derived from the drawings produced by the Design Department. Perspective pictures of the product are produced easily from solid models and with a little effort from 3D wire-frame models. Parts lists extracted from drawings into a computer file can be fed directly into the word-processor used for writing the manual.

It is useful for the Design Department to include maintenance data in the CAD drawing. By putting it on a special layer it can be plotted out specially for the Maintenance Department even using special colours.

Personnel

The Personnel Department will be impacted by the arrival of CAD/CAM in a number of ways. CAD/CAM introduces the requirement for new skills and new knowledge in the Engineering Department with the result that staff may negotiate for higher salaries on this basis. Also, management may ask the designers to work a shift system in order to increase the utilisation of the expensive capital equipment, which will also lead to new terms and conditions. In addition, the use of screens may raise new health and safety issues.

All the above affect terms and conditions but the biggest impact of CAD/CAM on Personnel matters is the need which it creates for expensive training. CAD/CAM training tends to be underestimated but is essential if the promised benefits are to be obtained. Heavy expenditure on training will be needed when the installation is first made. Furthermore, training will need to continue since every new designer recruited will need training. Fortunately, the cost can be considerably reduced by carrying out the training in the Engineering Department. Nevertheless, CAD/CAM is an activity which demands more training than others.

Finance

To the Finance Department the CAD/CAM system represents a capital investment in a new area of the company. Its benefits are different: they lie not so much in the labour displacement of electronic data processing (EDP) where there is a clearly defined short-term payback but in more intangible long-term things like design quality, reduced lead time or higher quality proposals. Is payback an appropriate way of assessing the investment in CAD/CAM? Also, a significant part of the cost of CAD/CAM is software which does not wear out as manufacturing plant does. Over what term should you depreciate it? Whereas EDP performs work that financial people understand, CAD/CAM uses computers to calculate geometry, which is far removed from their experience. All of this presents plenty of opportunity for misunderstanding which those promoting CAD/CAM should try to alleviate.

Data processing

As described in the previous section the CAD/CAM system is a radically different application of computers. Compared with financial applications it makes big demands on the CPU, it is interactive and cannot be fitted into a batch-processing schedule, and the files are relatively small with a structure known only to the software writers. It is an extremely complicated application requiring specialist knowledge to adjust it for efficient operation. However, it is still software running under an operating system, using disk storage space like the other applications and needing back-ups and a tape library. The relationship of the Data Processing (DP) Department to CAD/CAM will need careful thought.

If the CAD/CAM software is on a mainframe supporting other applications the role of the DP Department is fairly clear. It will ensure that adequate processing power is made available to what is a CPU-hungry application and provide data security procedures as with the other applications. It may find archiving requirements different due to the longevity of engineering data compared with its commercial counterpart. Since the CAD/CAM software meets the specialised needs of designers the DP staff cannot expect to support it; support will have to be provided by the CAD/CAM Manager who will be the link between Engineering and Data Processing and will probably report to the Design Manager.

If, on the other hand, the CAD/CAM system runs on a special computer located separately from the DP installation it will probably be managed entirely by the CAD/CAM Manager who will be responsible to the Engineering Manager. It would nevertheless be a wise precaution for the DP Department to be involved in setting up the CAD/CAM system and to allow them to give instruction on the basics of DP management, particularly with regard to data security and tape management. Common purchase and stock holding of consumables such as disks, diskettes and tapes would also be an efficient practice. In any case, it is highly likely that the CAD/CAM system will need to pass data to applications, such as materials requirement planning (MRP), computer-aided production planning (CAPP), stock control etc as mentioned in earlier sections of this chapter. Integration of file formats and communications links between the two computer systems will therefore be necessary and close cooperation between the respective departments essential.

Middle management

Of all the people affected by CAD/CAM it is the middle managers who may well feel the pain the most. Consider the manager of a department which uses CAD/CAM in some way. He may not have received any training because he is not actually going to use it, yet his staff will have been trained, will be doing their work in a way which is unfamiliar to the manager and using jargon that he does not understand. His confidence as a manager may well have been built on his superior experience of the work of the department as a more senior person. The arrival of CAD/CAM puts him at a disadvantage.

As described earlier, CAD/CAM offers the opportunity for better transmission of information between departments. The managers may feel that the power they have as a result of possessing information is being eroded. To take advantage of the better data transfer they will have to cooperate more with each other, in say, altering the format or content of the information passed on to another department.

So much for the problems: the opportunities are as great! Middle managers can make the introduction a success by motivating their staff, by being creative in looking for new ways of exploiting the technique, by making well thought out medium-term plans for its use and by making organisational changes to improve efficiency. They will need to manage the transition from a fully manual system to a fully computerised system with a policy for deciding which work is done manually and which on CAD/CAM, and they will need to devise and enforce new procedures.

The education of middle management in the technical and managerial characteristics of the new technique is paramount for successful application of CAD/CAM. Middle managers do not need to know all the ways of constructing a circular arc but they need to know that the arc is held as a series of accurate floating-point numbers. They do not need to know the commands in the operating system but they do need to know how drawings are identified, where they are stored and backed up, and how they are archived. They do not need to know FORTRAN but they need to know the precise nature of the data that is stored in the CAD drawing because it is this data which can be extracted and passed between departments to improve the efficiency of the company. They do not need to know how to design on the system but they must be able to sign on to the system, bring up a drawing and find their way around it. A middle manager who can do a few simple things on the system without embarrassment will undoubtedly earn the respect of his staff.

Senior management

High technology is always impressive but it still has to be operated by people, and many of the problems encountered will be people problems at root. Senior managers will not only have to adapt to the new methods themselves but ensure that their staff adapt as well: in particular their middle management. Senior management will need to encourage cooperation between departments to make the most of the opportunities presented for improved data sharing and transmission between departments.

CAD/CAM is not like hiring contract designers - a means of solving a short-term problem. Because it permanently changes the way people work it has a long-term strategic nature. It needs long-term goals and long-term plans to achieve them which only senior management is equipped to provide.

As discussed elsewhere in this chapter the introduction of CAD/CAM requires more than just the capital investment. It requires extensive training and the acceptance of a period of relative inefficiency while staff adapt. It is only senior management that can make such resources available in a way that does not adversely affect the company as a whole.

During the whole process of change there will be considerable problems, doubts and attacks of cold feet. To get the company through there needs to be a champion for the cause of CAD/CAM who is influential among the ranks of senior management. He needs to understand the company and CAD/CAM equally well, and be determined to see the project through against all resistance.

CIM: CAD/CAM AND INTEGRATION

Over the last 20 years the use of computers has spread beyond the financial systems into Design, Production and Management. While some companies have used one single computer for all applications, others have installed different computers in each area - probably because computers are marketed for particular applications and their reducing cost has made it possible to justify them against the budgets of individual departments. It is now becoming apparent that although the information used by the various departments is in computer files, it is often transmitted from one department to another by humans reading the printout from one computer and keyboarding into another, instead of being transmitted directly from one computer to another. The term CIM (computer integrated manufacturing or computer integrated management according to viewpoint) has been coined to draw attention to this state of affairs. It focusses on the benefits of connecting departments together with computers and on the technical and managerial steps needed to establish the connections.

There is an apocryphal story about the British bombing of German factories during the Second World War. When the Germans were questioned after the war about the effectiveness of the bombing it was discovered that the factories became more efficient as a result! The bombs destroyed all the paperwork and broke down all the internal walls. The result was staff could see each other and they started to communicate directly face to face or by telephone.

The Western world is being shaken by the overwhelming industrial success of the Japanese. Western pundits examine Japanese methods and bring back various management techniques - JIT, KANBAN etc. A Western journalist and his wife living in Japan and sending their children to the local infants and primary schools describes (Reference (30)) how the children are required to walk together to school in groups of about eight. The oldest child in the group, who carries a flag, is held responsible for getting the others to school. Socialisation is the central purpose of education. The individual use of a car is subordinated to the higher interest of getting a group to function smoothly towards a common purpose. It may be just that the Japanese culture encourages people to pull together and cooperate towards a single purpose.

Concurrent engineering

A product requires the contribution of many specialists such as design, production, packaging etc. Traditionally, each specialist completed his contribution before passing it on to the next one. The better information distribution methods offered by computer technology make it possible for the various specialists to contribute without waiting for the previous person to finish. The resulting reduction in lead time can be turned to advantage. The method is known as *concurrent engineering* or *simultaneous engineering* and requires new management structures as well as intelligent exploitation of computer technology. The contribution CAD can make is clear.

Information flows

Many of the opportunities described in the previous sections of this chapter arise because the design in a CAD/CAM system is in a computer file which can be read by other computers or programs and converted into data usable by other departments, provided the connections have been made. Table 25.1 shows the information flow to and from the Design Department and the type of data and software involved. The term "Customer" means the customer of design, that is, those initiating and specifying the design.

This can be Marketing, Sales or a customer of the company depending on circumstances. The variety of information is considerable.

Table 25.1 Information flow in design

	Information	Type	Software
	From Customer		
1	Description	Text	Word-processor
2	Desired performance and dimensions	Numbers	Database, Word-processor
	Feedback to Customer		
3	Appearance	Geometry	CAD
4	Predicted performance and dimensions	Numbers	Database
5	Predicted cost	Numbers	Database Spreadsheet
	To Production		
6	Geometry	Geometry	CAD/CAM
7	Parts and materials	Text	Database MRP
	Feedback from Production		
8	Preferred components and materials	Text	Database MRP
9	Preferred production tools and methods	Text	Database CAPP
10	Production costs	Numbers	Database CAPP
	To Buying		
11	Specifications, parts and materials	Text	Database, Word-processor
	To Customer Services, Technical Publications etc		
12	Drawings	Geometry	CAD
13	Specifications, parts and materials	Text	Word-processor
	To Costing		
14	Specifications, parts and materials	Text	Database, Word-processor

CAD/CAM FEATURES AND BENEFITS SUMMARISED

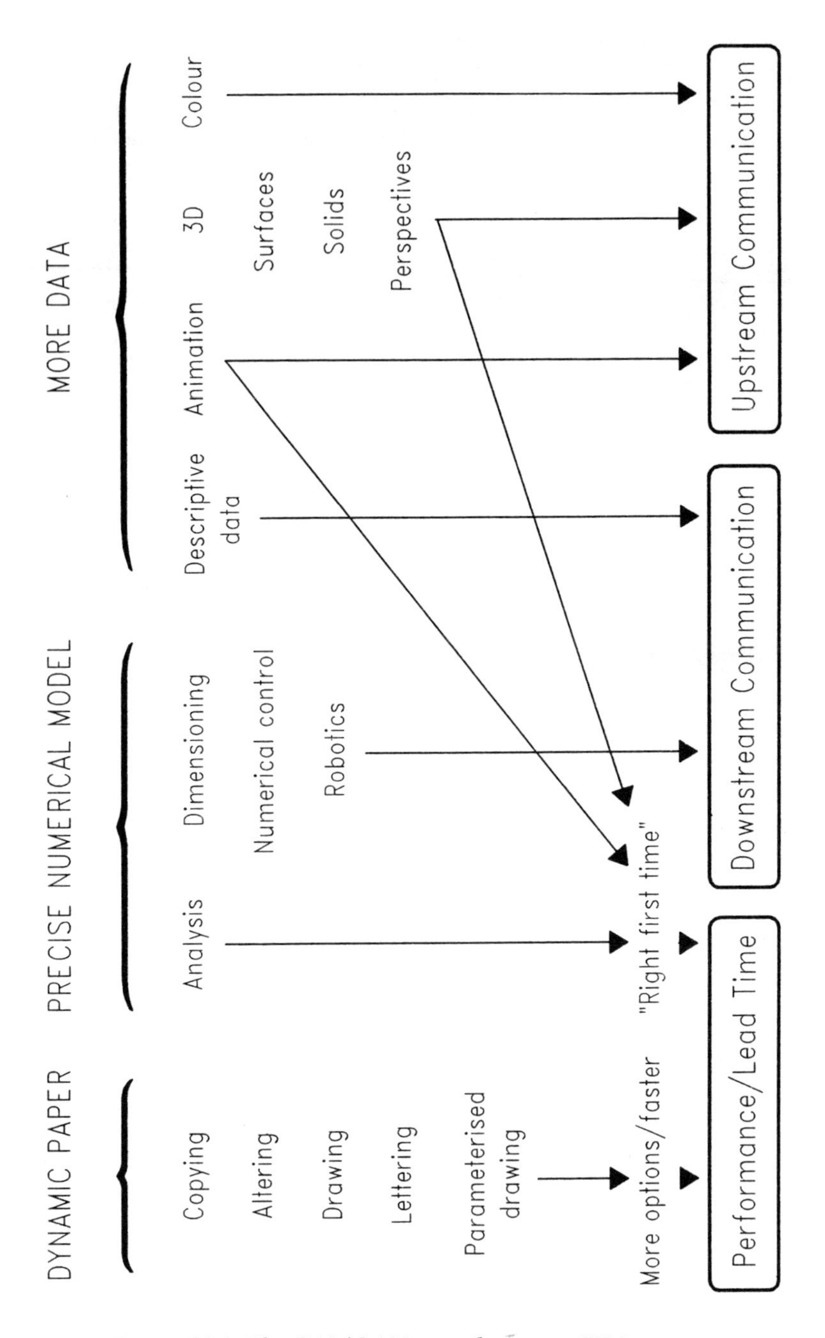

Figure 25.1 The CAD/CAM contribution to CIM

At this stage in the book it is appropriate to take a bird's eye view of CAD/CAM, its features and their contribution to the company as a whole. This is presented diagrammatically in Figure 25.1. Reviewing the range of technical features described in previous chapters we can summarise them into the three fundamental characteristics of a CAD drawing shown at the top of the figure. These are as follows:

Dynamic paper: Lines and text can be created, copied, repositioned, expanded and contracted with ease. The display can be magnified or reduced at will.

Precise numerical model: The display is the graphical representation of a mathematical model of the geometry held in digital numbers to high precision. This model can be analysed and dimensions calculated from it.

More data: The drawing is in a computer file and there is no practical limit to the type or amount of data it can hold. By selecting what is presented on the display or on the printout for the purpose at the time, confusion is avoided. The computer file can therefore contain a complete description of the design in every respect including textual information such as specifications, part numbers, finishes etc as well as the geometry to a high level of precision.

Each main characteristic is manifested in particular features as shown in the figure.

CAD/CAM offers three main categories of benefit. These, shown at the bottom of the figure, are as follows:

Performance/lead time: Firstly, CAD allows designers to design products with a better performance in the time allocated by letting them explore more design options, or it allows the time allocated for design to be shortened. There is always a trade-off between the performance of the product and the time allowed to design it. Note that this particular benefit has not been expressed as "labour saving" or "productivity enhancement". Improvements in product performance or faster response to customer requirements have a bigger financial effect than head count reduction since they are capable of increasing market share. Head count reduction in the Design Department has a relatively small effect on a manufacturing company where most of the costs are incurred in production or distribution. A better design, on the other hand, can increase market share by:

- Better performance or features than the competition

- Cheaper manufacturing costs than the competition

A shorter lead time can increase profits by:

- Introducing new technology before the competition
- Gaining contracts in the capital goods business

Downstream communication: CAD/CAM provides better quality data to the Production Department either as more consistent drawings without errors, clashes or parts which cannot be assembled, or as machining data for use directly by machine tools. The consistency comes from the precise model that the designer is creating so that clashes can be detected by analysis, and dimensioning is calculated from the actual dimensions of the model. The machining data is likewise obtained by extracting dimensions from the numerical model.

Upstream communication: CAD assists in the presentation of information to the customers of the Design Department, whether they are the company management or the customers of the company, by visualising the product in perspective projections, colour-shaded pictures or even animated pictures. They are able to understand the design better and are given confidence in it, whereas previously all the Design Department could offer was engineering drawings which only a trained engineer could interpret.

Chapter 26 Specifying and choosing the best system

Specifying and choosing a system involves considering a large number of complicated factors, balancing one against another and making judgements. In this chapter we shall try to get some kind of order into the confusing array of factors so that informed decisions and sensible judgements can be made.

We will start with the benefits which the system must achieve in order to justify itself. These are discussed in "CAD/CAM and its value" on page 13. For the purpose of this chapter we will combine the design labour and lead time reduction under one heading and add another factor to be considered: that of system management overhead. The objectives to be met are thus:

- Improve the quality of the design
- Reduce the labour or time taken in producing the design
- Minimise the labour overhead in managing the system

Each of these depends on particular characteristics - both positive and negative - of the system for their successful achievement. The quality of the design depends on:

- The completeness with which the product is defined in the CAD drawing or model
- The extent of the facilities for analysing the design
- The speed with which alternative designs can be explored

All of these factors are largely dependent on the range of software functions available and the range of geometric and descriptive entities available. In other words, quality is largely dependent on functionality.

The average time needed to produce a design is:

- Reduced by fast response times

- Reduced by the use of parameterised drawing software
- Reduced by a good repertoire of interactive drawing functions
- Increased by breakdowns in the hardware
- Increased by bugs in the software: these waste the designer's time in correcting the errors introduced and in finding ways round the problem
- Increased by a bad user interface

The system management overhead is reduced by good software utility programs and increased by unreliable hardware.

Looking through the lists of characteristics we can group them under three headings:

Function	The range of entities handled and the range of operations performed by the software. The numerical precision.
Performance	Computer response time. Storage capacity.
Reliability	The length of time for which the system does what is expected of it.

We will discuss how to optimise each of these in the decisions that are made in specifying and selecting a system.

FUNCTION

The previous chapters have reviewed the many facilities that are offered in CAD/CAM systems. In specifying the requirements it is necessary to decide the particular facilities which would be best for your needs. All systems will have a wide range of functions but not all systems have all possible functions so it is important to choose a system which not only possesses but is strong in the particular ones you are going to use frequently. We will therefore discuss the appropriate functions for a variety of common applications.

Before going any further it is worth remarking that numerical precision should be given some consideration. In most software it is usually sufficiently accurate but as one of the benefits of CAD/CAM is the very accurate mathematical model of the product which is possible, it would be foolish to lose the benefit on account of mathematical inaccuracies arising from inadequate precision. The topic is discussed in "Encoding geometry" on page 75. A way of testing for precision is to carry out repeated rotations of an entity which amount to one or more complete rotations and measure the difference between the final and initial positions. As the data is held in floating-point format the precision should be assessed in terms of the number of significant figures (or percentage accuracy).

Subcontractors and service companies

Subcontractors and those providing design services to other companies have special considerations in selecting a CAD system since they will need to receive or supply CAD drawings to or from their customers, and compatibility will be an important issue. Until there are better solutions to the problem of CAD drawing transfers the safest path is to pick the same CAD system as one's most important customer. Failing that, one has to carry out tests to ensure that satisfactory IGES transfers are possible. In any case, even when the STEP standard is working well it is wise to ensure that one has a system that can handle the same repertoire of entities as that of one's customer. Another alternative is to arrange to use one's customer's system over a data communications link.

Mechanical engineering

If the components are all turned or have perpendicular and parallel faces in the main then a 2D system may be sufficient. Where complex shapes and assemblies requiring analysis of complex three-dimensional relationships are involved then a 3D wire-frame system can be used with advantage. If the complexity extends to castings and manifolds with complicated surfaces then a simple surface modeller will be a useful addition to the wire-frame system. The design of production machinery may well benefit from a good mechanism package allied to the wire-frame system. Wherever stressing or thermal analysis is important a finite element mesh generator should be provided.

Solid modelling is a very useful design tool as an alternative to the use of a 3D wire-frame system where the components are produced by machining and therefore have fairly sharp edges. It is also the only practicable way of modelling components with enclosed spaces where calculation of the volumetric properties is important. Note that volumes can be obtained from surface models provided the surfaces generated all join up to enclose a space. Solid modelling, on the other hand, generates a space which is enclosed automatically.

Specials

Where a special version of the product is often designed to a customer's requirement and it consists of variations in just a few dimensions then a parameterised drawing system should be used.

Consumer products and plastics

The attractive appearance demanded by consumer products requires good surface modelling for their design. Reproducing the surface in production tooling will require CAM software. Assessing the appearance will require colour-shaded picture generation with fine control over colour.

Parts listings and other production information

Parts list extraction software will be needed. For specialised interfaces with production software, a good FORTRAN subroutine library will be needed to access the CAD database.

CAM

As discussed in "Numerical control program generation" on page 157 it is essential to meet the requirements of the Production Department. If finding a single system which is ideal for both sets of activities proves to be a problem then two compatible systems can be sought instead. As was shown, the CAD system contributes the geometry of the finished part but the NC program for making the part contains much more than the geometry in the finished state. The CAD system can also provide useful facilities for visualising the cutter paths and, of course, for designing jigs and tools. There are various ways of meeting the needs of the two departments and these can be discussed and debated within the procurement team. Nevertheless, the following requirements at least should be met:

- The CAD system must be able to extract the relevant geometry and pass it to the NC software in a usable format.
- The Production Department must be able to make changes and additions to the machine-dependent parts of the NC program.
- The Production Department must have post-processors which use its machine tools efficiently.

Various part programming systems are available. The choice should take the following into account:

- Preferably a widely used language
- Capability for writing macros for coding parameterised parts
- Good match to production plant
- Good and convenient user dialogue for editing the part program
- Good computational facilities

But above all else the system must have a full set of post-processors which will make efficient use of the machine tools in use. There are three ways of meeting the need:

1. Tailor-made post-processors
2. Post-processor generators
3. General-purpose post-processors

The tailor-made post-processor is the ideal solution and should be adopted if at all possible. It will have been written to exploit the particular machine tool to the maximum efficiency. A good post-processor generator can give reasonable results. General-purpose post-processors do not deal with the individual features of the machine efficiently.

Another approach is to buy a complete package of NC machine and NC programming system from one supplier. Such packages are not compatible with other systems and have the result of tying the customer to the supplier.

External and internal proposals and sales

A company making capital goods will want to produce timely and well presented bids, accurately costed. A "new products" group in a large company will want to present its proposal clearly and attractively to management. A supplier to a larger company will need to present various alternative solutions clearly to its customer in developing a final proposal.

A solid modeller or a surface modeller coupled to a good colour-shaded picture generator provides attractive, realistic and accurate visualisations of the proposed design. The surface modeller will need a 3D system to support it. Some solid modellers, on the other hand, can be driven from a 2D system. Careful attention will need to be given to the means of reproducing the colour-shaded pictures. Is it sufficient to show the customer the picture on the screen? Are colour transparencies to be made etc?

Another visualisation aid is animation in demonstrating, for example, a new mechanism. For those not wishing to go to the expense of colour-shaded pictures, intelligent use of coloured pens in a good pen plotter can improve presentation.

Besides good presentation, rapid and accurate costing is often needed. Where the offering is made up of a particular configuration of standard parts, the parts can be held in a symbol library with costs attached to them as attributes. Suitable data extraction software can then produce a costing from the proposal drawing itself.

Technical publications

Useful facilities for technical publications are producing perspective projections from 3D wire-frame, surface or solid models, automatic hidden line removal and automatic exploded views. There should be a high quality link between the output of the CAD system and the desk-top publishing or document composition software, if used, of the Technical Publications Department. The data format of the transfer file between the two systems will need attention. It may be useful to give the technical illustrator access to a CAD workstation.

Besides illustrations, technical publications will need parts lists on computer files for incorporation into their manuals.

Control systems

Electrical, hydraulic and pneumatic control systems all have schematic diagrams in common. Although three-dimensional presentations in isometric form are used in some schematics the relationships are by no means geometrically precise and the technique is purely a visualisation aid. A 2D system is therefore adequate in many cases. One case where 3D modelling is very useful, however, is in tightly packed hydraulic pipework. A 3D system can assist in the design and a perspective projection with hidden lines removed can provide a very useful guide to the installer.

A feature of all schematics is the large amount of descriptive data, much of which is extensively cross-referenced. Special software to manage this data will be of considerable use. For example, in a ladder diagram for electrical control it is useful to be able to renumber the relays while maintaining the cross-referencing between each relay and its contacts.

A schematic lends itself well to automatic conversion into production data of various kinds such as parts lists and wiring lists, so software to perform these functions is an important part of any CAD system for control circuits.

Process plant

Process plant, particularly plant on a large scale, makes the biggest demand of all on CAD and for this reason a number of special packages have become available. A large plant incorporates machinery, electrical power distribution, electrical and pneumatic control, pipework, conveyors and civil engineering earthwork, reinforced concrete and structural steelwork. It is designed by a multi-disciplinary team producing logically separate but physically integrated parts of the plant. Each group must be constantly aware of the work of the other groups. A cable tray cannot be allowed to pass through the centre of a pipe, for example! A large amount of the data is descriptive. An essential feature is a large database containing a single description of the plant which can be accessed by all groups. See Reference (39) for a recent report on the application of CAD in process plant design.

Electronic design

Electronic design is supported by specialist systems designed to manage the heavily cross-referenced descriptive information and the links between the schematic and the printed circuit board. A decision will need to be made on what to provide in the way of design analysis software, whether digital or analogue. Important features to consider are the various outputs to production systems and machines.

System administration

Desirable facilities are reviewed in "System administration facilities" on page 213. Plotting facilities are described in "Screen handling and output facilities" on page 133. Obviously, the more assistance the software gives to the system manager the better. Out of all of them, support for incremental back-up is essential.

Construction industry

Some CAD systems are designed specially for architecture and civil engineering. Useful features unique to this area are the ability to accept land survey data and map data in digital form. Another important function is extracting quantities from a design. Visualisation using perspective projection, hidden line removal and colour-shaded pictures allows clients to appreciate the design. The solid modelling functions to support this can be simpler (less accuracy, no Boolean operations) than those used in mechanical engineering. To obtain a proper indication of the effect of light and shade requires capabilities beyond most colour-shaded picture generators currently available. Producing a "tour" round the interior of a building involves making a large number of pictures which takes a considerable time.

PERFORMANCE

Of course, performance is something you have to pay for. You cannot afford to use a system with inadequate performance but, on the other hand, you will not want to pay too much for performance you do not need. Unfortunately, it is not easy to assess.

In the case of those choosing the system purely and solely for the ability to do something they cannot do otherwise, performance is of no concern: selection becomes a matter of choosing the software which will do the job and then choosing the hardware which will support the software.

The most important aspect of performance is the response time, that is, the time the user has to wait for the computer to do the operation he has requested. The factors affecting this are discussed in "The computer configuration" on page 59. The only way to be sure of adequate response time is to test it. The response time is increased by the amount of data that has to be processed in the operation and, in the case of a time-shared computer, by the number of other users the computer has to service at the same time. It should normally be negligible. To assess the response time, therefore, one must set up test operations in which an average amount and a maximum amount of data is processed and check that it remains negligible in the first case and is not excessive in the second. Suitable candidates for large amounts of data handling are:

- Picking an entity with the cursor in a drawing with a large number of entities (the software has to test all the entities)
- Performing a rotation on a large number of entities at once
- Hidden line removal for a faceted solid with a large number of facets

In the case of time-shared computers it is important to conduct the test with the maximum number of workstations in use at the time.

The two software functions which affect response time most, because they involve passing through the whole drawing data, are redrawing the screen and picking a geometric element using the screen cursor. Since they are used frequently it is important that they are fast. Both can be made faster by good software design and by special processors in the display hardware. Also, a high resolution screen will not need to be redrawn so frequently because the drawing will not need to be zoomed or panned so much.

Turning now to the issue of storage capacity: how much should be provided? Some people are tempted to think that the disk is where all of their drawings are going to be stored but this is a mistake. Firstly, data on the disk is vulnerable to failure of the disk. Secondly, since your whole purpose in life is producing new designs which are not going to be thrown away the stock of CAD drawings will be continually increasing and the disk will always be filling up. Thirdly, once a drawing is issued, it cannot be changed except by issuing a new version. When the new version is issued, the old version is kept for reference. Most engineering drawings are kept for many years and they are kept purely for occasional reference. It is clear from this that it is inappropriate, unnecessarily expensive and actually hazardous to use disk as the sole storage medium for drawings. Most of the drawings will be archival and should be stored on an appropriate removable medium. The disk should be thought of as the "Work In Progress" store, designs passing through it on their way to the archival store. The disk capacity should therefore be estimated on the number of drawings likely to be in progress at any time.

To assess the capacity required you will need to know how much space a typical drawing of yours occupies. This can only be done by either inspecting some drawings actually on the system or getting an opportunity by some means or other to input a typical drawing. It is not easy to estimate the storage required accurately although it will be broadly proportional to the number of entities.

RELIABILITY

The factors affecting reliability are covered in "Robustness, reliability and support" on page 217. It is achieved by the use of duplicated or redundant items in the configuration (see "The computer configuration" on page 59) and by purchasing from a reputable supplier with a good maintenance organisation. The supplier can be assessed by asking existing customers. Obviously, software and hardware reliability need to be assessed separately. Also needing separate assessment are the plotters and other graphic output devices. Pen plotters tend to be promoted on their speed but reliability, ease of operation and setting up are just as important. An unreliable pen plotter can create considerable system management overhead although it may not immediately stop the use of the system. As before, they can be assessed by asking existing users. Robust construction and the minimum of moving parts are good indicators of reliability. Printers and plotters are described in "Printers and plotters" on page 47.

CONCLUSION

Summing up the discussion of this and the previous chapters we can suggest the following priorities for arriving at the configuration:

1. Set down the software functions which are adequate for your needs and restrict your choice to systems providing them.
2. Choose a well established removable media device, preferably industry standard magnetic tape.
3. Choose a plotter which has no more intelligence, is no faster and is no bigger than you really need but has a reputation for reliability.
4. Decide whether to use a distributed or centralised/clustered system on cost, performance and reliability.
5. Choose the fastest CPU.
6. If you have chosen a processor shared by several workstations ensure that there is sufficient main memory.
7. Choose the display with the highest resolution and the fastest redraw.

EXERCISES

1. Is response time a matter of MIPS? If not, why not?
2. How should you estimate the required disk capacity?

Chapter 27 Conducting the procurement

In this chapter we will review the programme needed to acquire a CAD/CAM system. The extent to which all the steps described will be needed depends on the size of the company. A very small design consultancy, say, of a few people would not have the resources to carry them all out nor would it be appropriate. However, the principles are the same whatever the size. Reference (28) deals with the matter in greater detail with particular reference to medium and large companies and the integration of CAD/CAM into the whole structure of the company.

TO INTEGRATE OR NOT

It will have become apparent from "Management change" on page 235 that as well as being a means of making designers more efficient, CAD can be a component of an overall strategy for integration. If that is to be the case then the early planning will have to be more thorough in order to take the whole company into account. But integration is not for everybody. It is irrelevant for companies below a certain size and there will be larger companies who will choose not to attempt it for one reason or another. It is clear that the first question to be answered is: "To integrate or not to integrate?" Depending on the answer the planning will be narrower or wider in its scope.

RESOURCES FOR THE OPERATION

To be a success the procurement of a CAD/CAM system requires the full backing and commitment of senior management. The commitment will be expressed by providing adequate resources for the whole operation. These come under three heads: not just the obvious one of the capital budget for buying the equipment. Labour and financial resources need to be provided for each of the following:

1. The procurement and its preparation

2. The cost of the equipment

3. The upheaval in introducing CAD/CAM

The procurement will use up the time of a small team in internal meetings, meetings with suppliers, writing a recommendation, visiting trade shows and possibly attending educational courses. The time, together with the associated expenses, will need to be properly accounted for. The upheaval was discussed in "Management change" on page 235 and is not an insignificant cost.

If resources are to be provided then they must be properly estimated and planned, and proper plans cannot be made without facts obtained from analysis.

SETTING THE OBJECTIVES

It seems almost too obvious to say that the starting point must be a statement which is as clear and as detailed as possible of the benefits the company hopes to achieve from the operation. This may have already been done in making the case for CAD/CAM or it may be that just a simple decision to look into the possibility has been made. Whichever way, the following should be decided before proceeding:

1. What types of benefit should be possible and sought after

2. An estimate of the degree of each benefit where quantification is possible

3. The departments which would make use of the equipment

These are fundamental policy decisions which will direct the subsequent course of the project. They will need a consideration of the business objectives to be achieved by the investment in CAD/CAM. Is it better to have product quality, shorter lead time for new products, lower manufacturing costs, better proposals or what? What level of return on investment is expected? How does it relate to the strategic objectives? In addition, a knowledge of the nature of CAD/CAM facilities available and some kind of analysis of the work of the Design Department is required, including the extent of rework and other problems with costs assigned to the various activities if possible. A complete analysis in detail of the information and data flow between Design and other departments and their relative geographical locations will also be needed if integration is the aim. The company's existing use of computers will also be relevant. Answering these questions in a large company demands an extensive review by a team of several people. In a small company, a few meetings of Managing Director,

Technical Director and Company Secretary may be sufficient. However it is done, these questions must be answered before proceeding.

THE PROCUREMENT PROGRAMME

Of course, budgets cannot be properly allocated without a programme but setting out a programme also focusses all concerned on the hard decisions that will have to be made. Some companies get into a kind of endless cycle of sales demonstrations with trips to trade shows and vendors premises to see the latest goody. Every year brings some new development and it is true that any equipment procured will be out of date within 12 months. But the object is to obtain certain specific benefits and not to sample the flavour of the month. A planned procurement programme will maintain a focus on the objectives and establish a professional and responsible approach.

The programme should extend beyond installation and switch-on through training and well into the period of regular use in order to monitor the results against the original objectives: not for the sake of a post-mortem but to decide any changes of course that may be necessary as soon as possible. Obviously, the programme will include key decision points at which the project can be abandoned or shelved.

THE PROCUREMENT TEAM

The work of the procurement team is to gather information, interpret it and present a recommendation to management. It may then, perhaps in a reconstituted form, manage the site preparation, installation, training and change-over to regular usage. Its composition could include the following:

1. Representatives of user departments

2. The CAD/CAM manager designate

3. DP specialist

4. Trade union representative

5. Production representative

6. Representatives from other units or Head Office

7. Outside consultant

If you can include the CAD/CAM manager designate on the team you will have someone who is going to have to live with the results of the decisions more than anyone else. He will be strongly motivated as a result! The value of the DP specialist is in his general knowledge of computers and software and of the implications of linking the system to the company's other computer systems. Of course, if the CAD/CAM software is to run on the central computer then he will be deeply involved. The trade union representative may be involved because of the change in job descriptions and working conditions entailed. The production representative is particularly important if real CAD/CAM is envisaged with the CAD system providing data for the generation of NC programs and the various production control and planning systems. The representatives from other units are important to ensure compatibility with CAD systems in those units or to promote company-wide standards. The outside consultant has the advantage of being outside the sectional interests of the other members of the team and should have a wider knowledge of CAD extending to other companies and industries.

THE SPECIFICATION

The first duty of the team will be to formulate a specification for the equipment. This should include the following:

1. Number of workstations/terminals initially
2. Anticipated number of workstations/terminals eventually
3. Size and type of plotter
4. Output device for colour-shaded pictures if required
5. Maximum number of drawings active at any time, that is, the number on disk storage at any time
6. Estimated volume of plotted output for determining the speed of the plotter
7. Physical location of workstations/terminals for networking requirements
8. Archiving and back-up device and medium
9. Broad classes of application, e.g. mechanical engineering technical illustration, proposals etc

10. Major features such as parameterised drawing, solid modelling, numerical control program generation etc

11. Within each class, any cases where specialised design software would be of value such as reinforced concrete design, sheet-metal development etc

12. Hardware and software required for special interfaces such as to an existing parts catalogue, a CAD system in another unit, the stock control system etc

13. Dimensioning styles to be supported

14. Access control and document management facilities for issue control etc

In addition to the above you will need to add requirements for features peculiar to your taste or requirements such as particular symbol libraries, hatching patterns etc.

PRELIMINARY VENDOR SURVEY

Once a specification has been formulated the team can now collect sales brochures and visit trade shows. From the collection of sales brochures, a list of likely vendors apparently able to meet the requirements can be compiled. It is now necessary to go beneath the gloss of the brochures and start to find out the truth. The surest way is the benchmark, which will be discussed later, but benchmarks are time consuming and can really only be carried out on two or three systems as a final test of capability before selecting a system. You will need a way of reducing your list to two or three. A good way of finding out more is by meeting existing users. Another very effective way of getting at the truth is to read the user manuals. These have to describe what the system actually does although they are rather voluminous. If they are difficult to understand it will mean training problems later, or it may mean that the vendor has not properly documented or controlled the software development internally, both of which are warning signs.

For many companies who mean business in CAD/CAM the next step after the preliminary survey is to issue an invitation to tender to all the likely vendors.

THE INVITATION TO TENDER

The invitation to tender gives all recipients a clear statement of the requirements in writing together with useful background information, and specifies the format in which the proposal should be presented. It helps the vendors to offer the best system they can which meets your requirements and helps you make comparisons by getting them to present their proposals in the same format.

The sections might be:

1. Administration
   - Circulation of document
   - Time scale
   - Contact for questions
   - Evaluation method and other procedures
2. Background information
   - The work of the company
   - The size of the company
   - The objectives of the proposed system
   - Sites involved
3. Current computer equipment
4. The specification of requirements
5. Format of tender
   - Management summary
   - Hardware
   - Software
   - Support
   - Training
   - Implementation

- Contractual terms

PRODUCING THE SHORT LIST

The short list will be selected from those who respond with suitable tenders. While the vendors are preparing their tenders there will be the inevitable queries. Some procurement teams have chosen to answer queries at a single meeting of all interested vendors. In this way you can ensure that everybody gets the same story and the benefit of all the clarifications made, although the prospect of competing representatives all together in one room may be too daunting!

There is always the temptation to go to sales demonstrations. These can be time consuming, particularly as they usually involve visiting the supplier's offices. They do not contribute a lot to the procurement activity for the expense they entail and they are designed to conceal the defects of the system.

This is a good stage to do the all-important evaluation of the vendor himself. As discussed in "Robustness, reliability and support" on page 217, relationships with the supplier do not necessarily end when the equipment is installed. If a very popular software package running on a wide range of Personal Computers has been obtained then no further services may be needed from the supplier once the guarantee period has expired. New issues of the software with the new enhancements can be obtained from any other dealer. Widely available packages of this kind with a good reputation have usually had most of the bugs removed at a pre-release test stage. New issues therefore consist of enhancements. Hardware maintenance will be obtainable from any dealer.

The situation is different when one moves away from the high volume Personal Computer equipment into graphics workstations, mega-minicomputers and mainframes where the number of users is considerably less. It is possible for the level of bugs to be quite high so that bug-fixing issues are an important part of the after-sales service. Also, software houses supplying advanced high performance software in low volume are more likely to engage in continuous development with frequent enhancement issues.

If one has decided that the nature of the hardware and software is such that continuing support will be needed from the supplier then it is important to assess his capability both at the time of purchase and in the foreseeable future. Good signs are:

1. A large number of users

2. Vigorous commercial promotion

3. A healthy financial position

Bad signs are:

1. One, two or no users

2. Heavy reliance on technical novelty in promotion

The number of users is the most important factor. It indicates success and, should the worst happen and the supplier go out of business, it will be worth while for another company to pick up the product and trade with it.

THE BENCHMARK TEST

The short-listed vendors can be subjected to a benchmark test. The purpose of this is to prove the claims made and to discover problems. The software is so complex and the range of facilities offered so wide that it is easy to overlook vital aspects even after carefully drafting a specification. In the benchmark the candidate system encounters your particular way of doing design for the first time. Previously, all you may have seen are carefully rehearsed sales demonstrations designed to show off thc system to its best advantage and to gloss over deficiencies.

There are various ingredients that can be incorporated in the test and various ways of conducting it, but in all cases the vendor should be asked to reproduce drawings of yours. The drawings should contain idiosyncratic features - things that you do which are away from the normal run of design. The vendor is usually allowed to do the drawings in his own time and present the results at a meeting with you. However, an important ingredient is an unscheduled alteration to be carried out at the meeting without prior knowledge. Notice should be given that such a test will be made but the exact details not revealed until the meeting. The test will show up operations which, although they can be performed, take an excessively long time.

Other tests to consider making are a measurement of the storage used up by the test drawing and the response time to find a graphical entity in the test drawing. The effect of unexpected power failure should be discovered by asking to turn the power off while work is being done on the drawing. The amount of work lost as a result should be ascertained. Plotting speed may be another useful measurement. If a shared computer is used, response time measurements should be done on a system which is busy with other workstations at the same time.

It should be realised that the benchmark is expensive for the vendor to undertake so that he may refuse to do one. However, tests he will not be able to refuse are reproducing small portions of your drawings or other short unscheduled tests at a private sales demonstration. Under the same circumstances it should be possible to measure storage requirements and the speed of finding an entity in a large drawing, and to discover the effect of power failure. Large drawings can be quite quickly produced for these purposes by duplicating portions of geometry many times.

NEGOTIATING THE TERMS

After the benchmark, one or two vendors will emerge as satisfactory and terms can be negotiated. Up to the time of writing, price has been a very variable factor in CAD/CAM leaving considerable room for negotiation. Software pricing is particularly variable on low volume items as the supplier is largely trying to recoup the development cost over the sale of a small estimated number of copies. The cost directly associated with each copy is very small, being the cost of the medium, the manual and the installation work, if required. (It has been known for the price asked to halve after only preliminary discussions.) Sometimes software will be offered on a "test site" basis. Care needs to be taken to ensure that you know exactly what you are being offered for the quoted price as the range of alternatives and options with their complex terminology can be very confusing, particularly where a collection of programs is being offered under a single heading as a package. Software by its very nature is difficult to specify on account of its complexity. Other items which should not be overlooked are training arrangements and annual software and hardware maintenance fees, if relevant.

The situation is complicated if different parts of the installation are being obtained from different vendors since each part has to work with the others. For instance, the plotter may be supplied by a different vendor to the rest of the installation. Who takes responsibility for the equipment working together and how is this covered in the contract?

As the reliability and availability of the system are vital the choice of maintenance terms is an important consideration. The various maintenance facilities available are discussed in "Robustness, reliability and support" on page 217.

THE DECISION

Eventually, one supplier has to be chosen. It may be obvious after the benchmarks which one to choose. If the choice is difficult, one helpful technique is to set out a list of features or characteristics desired and apply a weight to each in proportion to its desirability. The weights of the characteristics offered by each supplier are then added up. If, in addition, the highest scoring supplier is also the one that is preferred from a "gut feel" point of view, confidence is gained for the decision.

Another evaluation, suggested in Reference (28), is to investigate how each candidate system would be installed, brought up to full use and operated in the first year.

The technical features are not the only factors which should be considered. Nor are they the most important. A number of very different aspects have been discussed in the foregoing pages. They are summarised below in the recommended order of importance:

1. Standing and future of supplier
2. Speed performance of hardware and software
3. Reliability and inherent robustness of hardware and software
4. Degree of match of facilities to requirements
5. Price

The standing of the supplier has a bearing on the quality of the product and is very important if support will be required in the future. The speed and performance is important because that directly affects the benefits to be obtained. Reliability and robustness are vital because failures can completely wipe out any productivity or speed benefits and actually slow down the users beyond manual methods.

EXERCISES

1. Do you need to see any standard sales demonstrations at all? Why not?
2. What can you do if a vendor refuses to do a formal benchmark?

Chapter 28 Site planning

Even setting up a new office requires planning to decide the best arrangement of furniture and a CAD/CAM system is considerably more complex. The following aspects will all need consideration:

- Cabling
- The location of the workstations
- Lighting
- Furniture
- Air cooling
- Fire detection
- Tape and disk storage
- Dust control
- The location of the central shared equipment such as computer, disk store, magnetic tape unit and plotter

CABLING

The arrangement of the site to accommodate the equipment will depend on the type of system to be installed which could be one of the following:

A A number of separate Personal Computers

B A number of workstations networked together

C A cluster of workstations sharing a minicomputer

D Workstations connected to a single large centralised computer (i.e. a mainframe)

In all cases, mains outlets need to be provided in sufficient number in the right locations to avoid untidy and dangerous mains leads strewing the floor. In cases B, C and D, signal cables to each workstation will be needed as well. Cables need to be kept out of the way both to protect them from damage and to prevent people from tripping over them. There are several schemes for doing this:

1. Suspended floors: This is the traditional way of handling cables in computer rooms. It is very effective and convenient but expensive.

2. Wall and partition trunking: This is convenient and easily altered but "island" workstations in the middle of the floor are not possible without some floor trunking as well. Floor trunking is not easily altered.

3. Wiring in suspended ceilings: This is not so common but is easy to install and alter. Some may find the dangling cables unsightly.

Cable identification

Wherever the cables are placed their management is a serious problem which is usually underestimated and sometimes ignored until it is too late. The problem arises because:

1. Even small installations require a large number of cables.

2. Cables need to be kept out of reach.

3. Once a cable is tucked away out of reach it is impossible to find its two ends unless they have been uniquely marked with permanent identifiers.

4. Cables are changed far more frequently than anyone could imagine.

5. You cannot change any cable whose ends cannot be found.

6. It is very difficult or impossible to find the ends of a cable without pulling it out which involves putting it out of service, and if the ends are not identifiable the effect of disconnection is unknown so it cannot be put out of service!

Do not lay a single cable until you have set up a cable identification system involving a scheme of identifying codes and a file in which each cable is recorded. Do not lay any cable without marking its separate ends - *even when you are short of time.*

Types of cable

Several networking systems are now available and more are likely to be introduced as the use of local area networks becomes commoner. They differ in the way the signals are transmitted and therefore in the type of cable required. The electrical signals reach very high frequencies: some as high as about 20 MHz. This is the same as radio and television transmissions, so cables are quite capable of acting as little transmitter or receiver aerials. Precautions have to be taken to ensure that the signal entering the terminal from the cable is what was put into the cable at the other end and not something it has picked up en route. The principal precaution is *screening*. The wires in the cable are surrounded by a flexible copper sheath woven of fine wire just under the outer insulation. This is connected to earth and radiation from outside is unable to penetrate it.

Another problem of conducting high frequency signals along a wire is that they are distorted by the proximity of other conductors close by. Complete compensation for the distortion can be made in the associated circuits provided the proximity of the other conductors is controlled. A conducting sheath provides such a tightly controlled environment for the high frequency signals inside. If the wire transmitting the signal is the only conductor inside the sheath then its environment is completely controlled and very high frequencies are possible. Such a cable is termed a *coaxial* cable or "coax". The higher the frequency the greater the loss or attenuation with distance along the cable. The effect is reduced by making the coaxial cable thicker but this makes it stiffer and more difficult to fit into cable ducts so there has to be some compromise.

When high frequency signals are sent along a coaxial cable the end of the cable must be terminated with the correct electrical resistance. If it is not correct the end behaves like an electrical mirror and signals are reflected back down the cable again, interfering with the arriving signals as they go. Normally, this is done by the equipment the cable is plugged into so that the user remains unaware of the need but in some types of equipment the cable supplies a series of terminals which just tap into it. Since only the person connecting up the terminals knows in this case which is at the end of the cable the equipment designer provides a separate terminating plug for the end. The configuration may work without this terminator but can be unreliable, so it is important to fit it when indicated by the installation instructions. The IBM graphics workstations (5080s etc) work in this way.

Every electrical signal must have a return path. Sometimes this is the earth or the screen. An alternative is to provide a second conductor specially for the purpose. If the two conductors are twisted together the chance of picking up radiated signals from other sources is reduced. The chance is further reduced if the pair is also screened. This configuration is often termed a *twisted pair* or *screened pair*.

A really reliable way of ensuring that a long cable does not pick up other signals or radiate its own signal is to use optical fibres in which the signal is transmitted as pulses of light down a thin glass or plastic fibre. Their use is becoming commoner although they are currently not as easy to connect together as electrical cables.

A useful product is the IBM Cabling System which can be used for local area networks and connections from terminals to mainframes. It includes a range of cables all comprising two coaxial cables in the same sheath plus connectors and some software to plan the layout of the cables and create identifying labels.

Signal cable topology

The paths of the signal cables to the terminals or workstations can follow either a *star topology* in which a separate cable is laid from some central point to each workstation or a *bus topology* in which cables are laid between terminals. Different networking systems also have particular *electrical* topologies. The signals may be distributed from a central point or may pass through each terminal in turn. But the physical cable paths need not reflect the electrical topology. Even if the signals are passed from terminal to terminal it may nevertheless be desirable to take the cables back through some central patch panel or marshalling point on their way from one terminal to another.

The topology to use depends on individual circumstances: the location of the terminals and the type of networking system employed. Certain networking systems require a star topology: asynchronous serial links, the IBM token ring and the IBM graphics workstations (5080 series and their successors) in particular. The IBM graphics workstations use the same networking methods as the 3270 series of terminals where the signals are distributed from a central controller with a separate coaxial cable for each terminal. However, the protocols allow several terminals to share a single cable. Each workstation provides an output for another workstation so that a limited bus topology can be created at the end of each spoke of the star. Other networking methods, such as Ethernet, have the electrical topology of a bus and others of a ring (in which the ends of the bus are joined together). For these the choice between a physical bus or physical star is a matter of judgement.

The patch panel

Whatever the networking system, a patch panel is strongly recommended. It allows rapid changes of connections to be made firstly to diagnose sources of trouble and then to isolate them. All connections, whether temporary or permanent, are visible and traceable so that the state of the connections can be determined at any time by inspection. A patch panel consists of an array of identical connectors laid out on a vertical panel which is easily accessible. Every cable in the system ends on one of these connectors which are also clearly identified. A set of identical patch leads then connects the ends of the cables in any desired configuration.

Asynchronous serial links

As discussed in "Hardware data exchange standards" on page 203, what was originally a means of connecting a modem to a computer or terminal has become a means of connecting a terminal to a computer. Such connections are used extensively to connect CAD workstations, terminals, plotters and printers to the computer, particularly in the case of a mini-computer or mainframe supporting many terminals. Only three of the specified RS232 circuits are needed: Received Data, Transmitted Data and Signal Ground. These handle the transmissions in each direction. The various "ready" signals are not needed. Flow control is handled by transmitting special control characters to stop and start the transmission. The connection can therefore be handled by a cable with three conductors, preferably screened. The frequencies involved do not demand the use of coaxial cable.

If only three conductors are used there is no need for the standard 25-way "D" connector at the patch panel. They are fiddly to plug in and wire up and have thin pins which can get bent. A good alternative is a robust version of the DIN connector with a "bayonet" type of locking ring for quick disconnection. You or your cabling contractor may have some other preference. Having said that, only three wires are needed for terminals, but what is needed for actual modems? This depends on the type of line in use. The control signals are needed where the line is a public dial-up one but this is rare for CAD since the sessions at the workstations are too long for public-switched calls to be economical. It is more usual to install private wires providing a permanent connection so that control signals may not be needed. In interactive operation, any failure on the part of the modem will be immediately detected by the interactive user at the terminal and there is no danger of data loss. For other types of operation, more of the control signals will be needed. It is useful to connect the modems through the patch panel, so an appropriate set of signals will have to be brought up to the patch panel for the sake of the modems even though they may be discarded for the local workstations.

The main problem in using asynchronous serial RS232 connections is the one described in "Hardware data exchange standards", namely the fact that the equipment on one of the ends of the connection has to pretend to be a modem. It is important to be consistent on which and decide either that the computer is to play the part of the modem or that the equipment connected to the computer is to play that part. If you are going to have a number of actual modems connected to the computer then you may like to make all the terminals, workstations and printers behave like modems. The difference is solely which of pins 2 and 3 of the RS232 connector conveys the data going into the equipment and which the data going out. Along with this you will need to decide what the pin allocation is at the patch panel. Somewhere along the route through the patch panel from computer to terminal a cross-over will have to occur so that pin 3 at the computer is connected to pin 2 at the terminal and vice versa. It can be placed on the computer side of the patch panel, in the patch leads or on the terminal side of the patch panel.

In general, computers can send signals over asynchronous links to terminals located in the same building. The exact limit on the length of cable depends on the actual driver card in the computer and should be ascertained from the supplier. To drive longer lengths of cable, typically on a large factory site specially designed, relatively cheap units, usually called *line drivers*, are available. They can be used on wires provided by British Telecom for internal telephone systems with suitable permission or on your own installed cables. For terminals situated outside the site then the proper modem comes into play.

Power distribution and control

The installation will comprise quite a large number of separate electronic units such as terminals, workstations, computers, printers and so on. Each will require a separate mains outlet. Although the total electrical load may not be particularly great, a large number of outlets will have to be installed.

If the minicomputer in case C is to be used then it is convenient to run all the units in the computer off the same circuit (perhaps a single consumer unit). Operating a single switch ensures that the whole minicomputer and its associated units are off when required. This is particularly valuable in relation to fire precautions.

THE LOCATION OF THE WORKSTATIONS

Although some early systems placed a restriction on the length of cable between computer and workstation there should be no problem with current systems. The location and positions of the workstations can therefore be governed by human and management considerations. It has been common to have a special room for CAD workstations separate from the Design Office but is this the best way, bearing in mind the amount of informal communication which is essential in a Design Office? It arises from and reinforces the idea that CAD is separate from "real" design, is only for certain types of people or only for specialised work. In order to integrate the CAD work and assist in a smooth transition to CAD the workstations should be in the same place as the designer's desks and drawing boards (if still in use). Furthermore, the need for supervision will not change. It may actually need more supervision in the early stages in order to enforce the various new practices which arise as a result of using CAD. It is true that because workstations will have to be shared on account of their cost, it is not possible to position each one next to a designer's desk. Also, because a session at a workstation is more concentrated than that at a drawing board, discussions between designers should take place away from the workstation. The foregoing considerations suggest that the best position is in the same room, in a group together, away from the desks and drawing boards.

LIGHTING

The big difference between the graphics screen and the paper it replaces is its optical characteristics. Paper works by reflecting the incident light diffusely while a graphics screen works by emitting light but, being glass, reflects incident light like a mirror. All sources of light located behind the user appear superimposed by reflection on the picture he is trying to interpret. The best way of avoiding the problem is to position users with their backs to walls. Alternative solutions are using large area, low intensity, lighting such as illuminated ceilings or uplighters, or by installing special ceiling diffusers which do not emit light sideways. Dim lighting is unsuitable as users have to refer to papers etc while working. Desk lamps can also be picked up on the surface of the screens.

FURNITURE

The requirements for furniture were discussed in "Workstation layout" on page 43. The furniture supplied by the CAD/CAM vendor is unlikely to be adequate so suitable additions, such as angled reference tables, may have to be ordered.

COOLING

The large amount of electronic equipment will generate considerably more heat than is normal for offices. In Europe the extra heat will not normally be noticed until exceptionally hot days occur in summer. Under these conditions the extra heat is no longer conducted away by natural ventilation. Working conditions will become impossible for the humans and the computers alike. Equipment should be installed specifically designed to remove the excess heat under these conditions. It usually consists of wall-mounted air coolers.

FIRE DETECTION AND CONTROL

In the case of minicomputer installations it is usually found convenient to run the computer continuously round the clock. Experience of running electronic equipment shows that it is noticeably more reliable if left running, due probably to the lack of thermally induced stresses as it warms up and cools down. Continuous running also allows greater usage of workstations with a shift system. The problem with continuous running is the fire risk presented while unattended at night. To meet this, smoke and heat detectors should be installed if possible and connected to a fire alarm so placed that it can be acted upon by a night watchman or the fire brigade. The fire brigade will need to be shown how to switch off the power and suitable fire extinguishers will have to be installed and maintained.

TAPE AND DISK STORAGE

This subject is covered fully in "Data security and contingency planning" on page 299. The requirements are controlled humidity, protection against fire damage and possibly restricted access to staff. The storage area will have to be visited about once a day for the back-up procedure. (See "Back-ups" on page 300.) There are two solutions to fire protection. One is a fireproof safe near the central equipment and the other is a place in a separate building where duplicate copies are kept. In both instances the humidity will need attention to ensure long-term stability.

DUST CONTROL

As discussed in "The computer configuration" on page 59, even minute particles of dust can cause a disk head to scrape the recording medium and destroy the data stored there. For this reason it is important to reduce dust in the atmosphere of the room where a disk is working. Steps which can be taken are keeping the room closed, minimising the number of people entering it, installing special carpeting and arranging the air conditioning to blow filtered air into the room so that all air movement carries dust out of the room. The problem applies to disks which are regularly removed from their drives but not to permanently sealed disks, usually known as "hard" disks or Winchester disks.

THE LOCATION OF THE CENTRAL SHARED EQUIPMENT

Every type of configuration will have some central shared equipment. Even self-contained workstations networked together will probably use a file server and all installations need a plotter. Being shared, if any of these items fails it can bring the entire installation to a halt so that extra protection may be needed.

Central computer or file server

The main hazards are malicious or accidental interference by unauthorised or untrained people, fire and, in the case of unsealed disks, dust. The suppliers may specify strict environmental limits on temperature and humidity for reliable operation. All this points to locating the equipment in a separate room with controlled access by door locks and a controlled environment. The CAD manager will need access in order to supervise and control the system but it may not be necessary to locate it near the CAD manager's office as it can be controlled through a terminal in his office. Access to the office will then need to be controlled. An important requirement from the supplier may be for sufficient space around the equipment for maintenance access. This can be quite large as units have to be slid out of the cabinet for attention.

Plotter location

The location of the plotter will need a little thought. It will give off dust from the paper movement which means it is best kept separate from unsealed disk units. If it is a sheet-fed device, frequent access will be needed and the store cupboard of sheets will need to be near to it. If drafting pens are used then there will have to be a table close by to keep the pens, ink bottles and pen cleaner, and a sink or hand basin for cleaning pens not far away. If it is roll fed, a large table and a large paper trimmer also need to be close by to cut up the rolls after plotting. The location will depend on whether designers are going to attend to their own plots or whether the plotter is going to run unattended except for occasional visits by an operator.

EXERCISE

In what order do you think you should tackle the planning of the various aspects of the installation?

Chapter 29 Implementation

This covers a period starting with the commitment to procure a particular system. Implementation can be regarded as being over when the performance of the system has been reviewed against the original objectives and been considered satisfactory. Putting it in accountancy terms, implementation is the period of maximum negative cash flow. It will involve heavy expenditure and it will be a time during which the designers will work less efficiently than they did before. Clearly, everything must be done to reduce the ill effects. The more care given to planning the better, and it should be treated in the same way as any other project. Because there will be many who are secretly uneasy or even openly critical the policy should be to get some visible success as early as possible and to develop from that position.

The tasks to be done are:

1. Site preparation
2. Training and education
3. Installation and acceptance
4. Customisation
5. Cut-over
6. Review of achievements

MOTIVATION

The decision to procure the equipment involved motivating the senior management but so far it has been unnecessary to motivate anyone else beyond the team directly involved in the procurement. Now the designers are going to be hit hard by the change. They will have to learn radically new skills and gain at least some small understanding of computers. These are activities requiring extra effort and some discomfort for which encouragement will be needed. Fortunately, many designers by virtue of their job are interested in new things, which will be a help, but not enough in most cases to overcome the frustration of having one's designing slowed down by a new and difficult tool. To provide encouragement, emphasis can be placed on the way CAD makes better designs possible both by its analysis facilities and by the opportunities it offers for exploring many more alternatives in the available time.

Conversely, presenting CAD as a way of increasing productivity (which everybody knows means head count reduction!) could be very demotivating. Another discouraging factor could arise from staff experiencing pressure to meet stiff completion dates. Unless due allowance is made the CAD system will be seen as an extra handicap in a difficult situation. Another demotivating aspect will be experienced by staff with supervisory responsibilities who fear being upstaged or made to look foolish by younger junior staff picking up the new technique more quickly than themselves. This can be avoided by appropriate training and education.

TRAINING AND EDUCATION

Designers have to be brought up to a state of efficient use and management needs to develop confidence in the system and understand its capabilities sufficiently to exploit it to the full. All this involves a considerable amount of training and education which, because it cannot be done instantaneously, requires planning. What are the priorities? The highest priority is to get some designers using the system efficiently as management education will arise in part from seeing some successful projects. In training the designers the policy should be to aim for the greatest efficiency of use, albeit in a few users, rather than the greatest number being able to use it. It is fairly easy to find a few who will learn quickly and who can rapidly go on to produce visibly successful results. In so doing they will achieve several things. Their success will encourage the less able to tackle their training thoroughly, it will generate confidence generally and the experience they gain in developing the best procedures for the company's particular application of the system can be passed on. The latter point is important as efficient operation depends very much on knowing the best procedure for a particular design problem. The only danger in concentrating training on a few is that it encourages the idea that CAD is something special for a limited number of special people or a few particular applications. The ideal is to ensure that both senior and junior staff are included among the initial trainees. Training at least one experienced senior person indicates that the company is serious about CAD.

The training needs to be carefully scheduled. It is important to have the system in efficient use as soon as possible and doing training before installation at the supplier's or a training company's premises will help towards this. On the other hand, it is important to avoid the situation where trained users cannot immediately start using the system. As efficient use depends largely on practice, a user's efficiency disappears quickly if he does not use the system frequently. The skill learnt on the training course is lost if he does not immediately put it into practice, quite apart from the frustration which results. There may be a temptation to have a big course for everybody all at once at the beginning before regular usage has been properly established. Once again, a solution would be to train just a few before installation, as suggested above. A full discussion on training is given in "Training, manuals and user groups" on page 287.

SITE PREPARATION, INSTALLATION AND ACCEPTANCE

Site planning is discussed in "Site planning" on page 271. Since preparing the site will probably require alterations to buildings and services, it must be scheduled so as to be completed before the computer installation commences. The supplier or suppliers will deliver and install the equipment. The question arises: "What degree of acceptance test is appropriate for such a complex system?" For the reasons discussed under software reliability elsewhere in this book it is practically impossible to test out software completely. Instead, one has to take it as it is and trust the supplier to correct malfunctions in subsequent releases. A problem that can arise with some software suppliers who have not established good software issue control procedures is deciding if the package is complete, as it will consist of a large number of utility and ancillary programs and even one or two extra "goodies" thrown in for good measure besides the main program.

However, there is no reason why one should not test the hardware for proper operation. The computer should power up properly and be able to run the software. Computers usually perform diagnostic software automatically on powering up and often do it continually in spare time while they are running. The most likely sources of trouble are in the input/output devices, particularly the plotter. At least one essential acceptance test is to produce a drawing on the system and have it plotted out. The magnetic tape unit, if provided, should also be proved out by writing and reading a tape. The lines to all the workstations can also be tested.

Particular problems arise if more than one supplier is involved as they will blame each other if they possibly can. Plotters, for some reason, are particularly difficult for software to drive. The designers of plotters seem unable to provide a simple interface. Interfaces between one supplier's equipment and another's should be given particular attention. Where they fail to communicate, representatives from each supplier should be summoned to the site on the same day and asked to solve the problem together.

PREPARATORY DRAWING, CUSTOMISATION AND CUT-OVER

The company drawing frame and frequently used symbols are items which will be required right from the start. The schedule should include drawing these on the system before any other work is done. It may be an advantage to do the work on a bureau or by hiring the use of an established system before installation. In addition, there will be various standards, particularly in dimensioning, which will need programming into the software. These are the initial steps in a steady process of adapting the software to the practice of the Design Office. Unfortunately, this initial customisation, which will affect the future functioning of the system, has to be done when there is the least experience with the system.

One difficult question which has to be settled is what existing drawings should be redrawn as CAD drawings. The guiding principle here is that CAD only gives benefits in proportion to the extent to which the CAD drawings are used subsequently, whether by other departments or by the Design Department itself. When you put an old drawing into CAD the benefit you gain from doing so arises solely from the greater depth of information it holds and the ability of software to make use of that information. If that benefit cannot be realised then there is no point. Equipment does exist to scan old drawings but it should be clearly understood that such equipment rarely generates the depth or accuracy of information that a properly produced CAD drawing holds. After all, it cannot generate information that is not there in the first place. The point of CAD drawings is that they hold more data than a paper drawing.

When installation is complete there should be some designers already trained waiting to start work and a plan detailing what work is to be done on it. The work selected for this early phase will be determined by various factors but it is worth choosing work which will show off the system to its best advantage, bearing in mind the relative inexperience of the users. A good choice is one complete but small design. The biggest gains are obtained by doing a complete project on it although this will require greater courage.

Although the links with other departments and systems are important, implementing them is clearly best left until the CAD system is a going concern in the Design Department. However, their eventual use must be planned in detail at this stage since they will impose requirements on the standards, procedures and customisation which will be set up in the early stages. It is at this stage that decisions are made about part number and drawing number formats, what data is put into the drawing etc.

REVIEW

Since the procurement was agreed on the basis of achieving certain objectives there should be a review after about a year to see how well the system has lived up to its promise. There may be some objectives which have not been met but there might also be some unexpected benefits which have been discovered. Some applications of the system may have turned out to be particularly fruitful and others a waste of time. The review will therefore report to management on the degree to which the objectives have been met and make recommendations on how the system should be applied in the future. Who does the review depends on circumstances, whether it is performed by outside consultants in the case of a large installation in a big company or whether it is simply a discussion between the Chief Designer and the Technical Director for a two-workstation system in a small company. However it is done it ensures that any misjudgements made in the process of justification and procurement are corrected.

EXERCISE

At what point in the schedule do you think it would be best (if at all) to run an "awareness course" for management?

Chapter 30 Training, manuals and user groups

It is clear that training is essential for the efficient use of CAD: the technique is revolutionary and new to everyone. As a result the cost of training must be included in the budget together with the other costs of procuring and owning the system. Although the cost is highest at the beginning it will continue for as long as CAD/CAM is in use since new staff will need training when they arrive and new facilities provided by the CAD supplier will have to be learnt. How much resource is devoted to it will depend on individual circumstances so it is not possible to give any general guidance. However, one can indicate the consequences of not undertaking training. Even if a budget is allocated there is still the problem of finding time for training in a busy company. Since it does not produce immediate visible results it gets a low priority. It may be possible to link training with productivity by reporting gains appearing as a result of training. It can be done outside normal hours or when the work load is light. There are four types of training required which will be discussed in the following section.

TYPES OF TRAINING

User training for efficient design

This is the training for those who will use the system. The most important objective is to make the operation of the software second nature. Using the system should require as little thought as using a pencil since that is what it is replacing! The only way to reach this state is practice as one would in learning a foreign language: in fact the designers are learning the language used for telling the software what to do. Although vocabulary and grammar play their part in learning a foreign language, fluency can only be obtained by practice in actually using the language in realistic situations. It is not enough to teach what each command does, as has often been done. The sequence of commands required to achieve any particular result must be taught, rather as idiomatic phrases have to be learnt in a foreign language. From all this, it can be seen that an effective CAD training course must consist predominantly of exercises based on realistic bits of engineering drawing. One consequence of the importance of practice is that users can easily get out of practice if they do not use the system sufficiently frequently. Once the initial training has been done they need to use the system as soon as possible.

Seeing that practice is obtained on the actual job it could be argued that there is no need for formal training beyond that of some theoretical instruction. But the user has to be brought to the point where he feels he can use the system productively, if not efficiently, otherwise he will find it just an obstacle in his work. So well planned instruction coupled with graded exercises is needed. An important milestone is the point at which the designer becomes as efficient as he was with a pencil. Perhaps after this milestone, formal training is not absolutely necessary although even then there will be many facilities he is not using to advantage because he does not know about them. The consequence of not continuing training in some form or other is that inefficient practices will get established. Perhaps short periods of instruction or the use of computer-aided learning or other self-study methods will be sufficient to extend gradually the user's ability. Opportunities for users to share "tricks of the trade" for short cuts and efficient usage are of value here.

Supervisor training

Those managing designers using the CAD system do not need to be efficient users of the system but they need to be able to examine a CAD drawing in full detail in order to inspect and check the work. They should also have a thorough understanding of what the CAD system can and cannot do in order to be able to schedule work. A sound theoretical understanding of the structure of the CAD drawing and the functionality of the system will gain the respect of their expert CAD users. The consequences of not training supervisors is that they will not give wholehearted support for CAD to make it a success. Instead, they will pay it lip service while continuing to act as if nothing has happened.

System administration training

This is training given specifically to the CAD manager on using the many utilities provided for data management and security, user registration, accounting and customisation. It can and should take place before installation with perhaps a further course on the more advanced aspects some weeks after installation. Unfortunately, it still tends to be neglected by CAD suppliers. The worst consequence of not doing this training is that CAD data and drawings are lost because the data management has been done incorrectly.

Management awareness training

Awareness is really too weak a word in this context as it suggests something nice to have but not really essential, going with the idea that CAD is just a rather expensive item to make the designers work faster. Anyone in the company that has to make decisions from the Board downwards needs a clear understanding of what CAD/CAM can and cannot do for the whole company although they do not need to know how it does it. Aspects which should be covered are the data that can be put into the CAD model or drawing, the various ways in which this data can be used outside the Design Department, and how this can be realised in benefits of shorter lead time, higher productivity and higher quality. The consequences of not doing this training are that the full potential is not realised and opportunities are missed. It might even result in CAD being no more than an expensive toy. The training will largely consist of discussion although some simple "hands on" exercises will help to develop confidence and give insight.

SOURCES OF USER TRAINING

There are several potential sources of training although not all will be available for any particular CAD/CAM system:

1. Off-site courses run by the vendor
2. Off-site courses run by a specialist training company
3. Home-brewed training of the "sit by Nellie" variety
4. Manuals
5. Computer-aided learning
6. User dialogue

Off-site training

One of the constraints to be overcome in providing training is the heavy use it makes of the expensive resource of workstations. Since the objective is to develop fluency there is no way of avoiding their use. The problem is less apparent during initial implementation as the workstations are not being used productively anyway, but it is an issue in training new staff in an already efficient and productive installation. The only solution is to train staff off-site at vendor or training company courses.

"Home-brewed" training

Home-brewed training appears easy but is in fact very expensive. It uses up valuable workstation time and to do it efficiently requires generating one's own exercises. This involves considerable work as a large number of exercises have to be devised and then documented. On the other hand, you will develop practices particular to your own type of work in which staff will have to be trained, so a certain amount of in-house training will have to take place.

Self-instruction

The types of training listed in 4, 5 and 6 can largely be regarded as important supplementary sources after the main training has been used to develop the basic fluency.

Manuals

There are two ways of organising a manual, corresponding to the two approaches to training. In the better documented systems this leads to two parallel sets of manuals. The first kind of manual is always provided and consists of a dictionary of commands with a separate section on each command describing its action and all its parameters and options. The order of the commands in the manual follows some pattern related to the commands themselves, such as alphabetical by name or broad class of function. Since it can only be used by someone already reasonably familiar with the commands its main purpose is to extend the knowledge of a user who has already been trained by allowing him to study new options or parameters not previously covered.

The second type of manual is suitable for the untrained user. It is organised according to the operations with which a designer will already be familiar, such as "dimensioning" or "constructing angles", and describes the sequences of commands which can be used to achieve particular results. Combined with a good set of exercises such a manual can allow a user to teach himself provided there is someone on hand to help out occasionally should he get confused.

Computer-aided learning

A number of computer-aided learning (CAL) packages for CAD are now available. The advantage of CAL is that it allows the user to work at his own pace and is less intimidating for some people. A CAL package is also a once-off cost compared with the recurrent cost of training courses.

User dialogue

The user dialogue should be regarded as a training facility built into the software itself. At the time of writing there is plenty of room for the development of this aspect of CAD software which can considerably help to increase efficiency. Context-sensitive help software goes some way towards this end. The idea that the user dialogue is itself a training aid may be new to some but the principle is well established in human dialogue. If you are talking to someone and he clearly does not understand you it is natural to amplify what you are saying and even give an explanation. There is no reason in principle why software should not do the same. At the point where the human makes a mistake the software has all the information it needs to give precisely the right instruction. In the days when computers were very expensive machines devoted to repetitive calculations, a program getting an input which was wrong simply gave up after emitting a cryptic error message on the basis that it was the job of humans to supply the right data. Since a computer is now a tool for increasing the effectiveness of a human it should give every assistance possible. It should lessen the hold-up in the user's work when he makes a mistake by trying to discover what it is and correct it, much as another human would. Of course, handling an error condition by analysing it and engaging in further dialogue costs very much more programming effort than just simply aborting.

THE USER GROUP

When well run, a successful user group achieves formal and informal communication between different users and between users and the supplier. It can ensure that the supplier sets the right priorities in fixing bugs and developing new features. It is a useful source of experience and is worth contacting during the later stages of evaluating a new system. It can also be a source of symbol libraries and ancillary software. Often, a supplier will indicate the direction his development is taking and announce new features under consideration.

EXERCISE

Of the four types of training, pick the two most important:

Before installation

During the first year

After the first year

Say why.

Chapter 31 Efficient usage

In this chapter we consider aspects of management which are special to CAD/CAM systems and which need attention to get the most from the installation.

THE SELECTION OF WORK

Unless the bold decision has been made to switch over completely to CAD and abandon the drawing boards, there will be some work done on paper along with the work on the CAD system. This requires a decision to be made for each new drawing whether to do it on CAD or not. The temptation is to decide this literally on a drawing-by-drawing basis, simply allocating a new drawing to a CAD designer when he is ready for one. Following such a policy loses the advantage gained from CAD drawings being a complete and numerically precise definition of the product. The greatest benefit is obtained from this important characteristic if an entire project, product or machine is done on CAD from layout to detailing. The layout drawing is done in more detail than would be on paper. More time is taken over it and all parts are drawn accurately with full detail where they interact with other parts. In this way, clashes are avoided. Detailing is done by taking copies of the parts drawn on the layout and putting them on separate sheets. Because of the precision of the layout the drawings taken from it are precise. Detailing consists of putting in internal detail not covered by the layout, dimensioning up and providing full visible documentation of the component for manufacturing. Detailing is much quicker and labour saving is achieved because the parts are not drawn twice. Neither this nor the clash avoidance is possible if CAD is used in a piecemeal fashion on a drawing-by-drawing basis.

SCHEDULING WORKSTATIONS

If the procurement of the system has been justified on a productivity basis then a high utilisation of the workstations is important. This can only be achieved by scheduling their usage. Leaving the use of a workstation to chance will result in a low utilisation. Even providing one workstation per designer will result in a low utilisation since designers spend a significant amount of time in activities other than drawing.

One method often adopted for increasing the utilisation is to run shifts, a popular scheme being two shifts overlapping to allow for meetings and discussions.

STANDARDS

A CAD system introduces standards of its own in addition to the existing drawing office standards. A particular one is the assignment of layers. As some systems provide for a large number of layers it is quite easy for chaos to develop as a result of users putting things on different layers. The effect is that items go missing when a drawing is plotted. Unfortunately, the assignment of layers needs some experience of using the system in the particular Design Office. However, discipline can still be exercised right from the beginning by making some of the more obvious assignments and enforcing them either by instructing users and checking their work or by customising menus appropriately. The general principle is to use layers for controlling how various classes of item shall appear - when they shall be visible and when hidden. Some systems use layers to control the colour or the plotter pen to be used.

Another area where discipline may be required is in the choice of text sizes and styles, particularly for systems which offer a wide variety of fancy styles. Plotters can produce legible text down to about 1.5 mm in height. However, subsequent reproduction processes applied to the plotted drawing, such as microfilming, may make it illegible. This will need to be thought through early on and a standard established and applied by appropriate customisation of the software.

Considerable productivity is obtained by the use of symbols so that the early development of a symbol library is valuable. The library is a set of standards so that control over alterations to it is important. Some of the more enthusiastic designers will want to be continually making additions to it while others will be tempted to create private libraries of their own in conflict with the common library. Some systems provide symbol facilities where changes are retro-active or can be made so. In other words, the plotter produces the shape of a symbol as it is at the time of plotting and not as it was at the time of drawing. This can be an advantage or a serious embarrassment depending on circumstances. The issue needs to be thought through early on and a standard adopted.

Dimensioning styles are another standard which will already be established before CAD. They should be customised into the software from the beginning.

DRAWING MANAGEMENT

Up until the year of writing, CAD systems have sadly neglected the potential of computers to handle the management of drawings. In fact, some systems have made it more difficult. They have used the file identification methods provided by the operating system for storing drawings and this has limited identifiers to hopelessly short lengths of eight characters. In addition, the facility for creating and altering copies of drawings opens up the serious danger of multiple masters. As soon as two people simultaneously modify two separate copies of the same drawing there is real trouble since the one drawing has turned into two different drawings which will take a lot of work to combine back into one again.

Drawing management software linked to the CAD system is becoming more common. The ideal facility should allow all the information usually held in the drawing border to be recorded in a database and the drawing to be selected and retrieved using the information. It should also control work on the drawing to prevent two people working on it at the same time, and handle the issue process and the raising of modifications, producing any related documentation.

PEOPLE PROBLEMS

The introduction of the CAD system can create some people problems which reduce the efficiency of operation. The CAD/CAM or design office manager should be aware of the various kinds of problem so he can spot them when they occur and deal with them.

Apart from the danger of generally unenthusiastic designers a potentially serious problem is the manager or other influential person who is opposed to the new system but has not so far revealed his true feelings. This is best avoided by ensuring commitment and open discussion from all concerned during the whole procurement activity.

Fairly easy to deal with is the designer who just finds it difficult to adjust to using CAD. His problem will soon become apparent unless the supervision is completely inadequate. The solutions are either giving extra training or finding other work for him.

The other people problems peculiar to CAD systems arise from the opposite of negative reaction - enthusiasm, but enthusiasm going off in the wrong direction due to the addictive attraction computers have for some. One direction for misplaced enthusiasm is in creating excessive detail in the drawing simply because the software makes it possible. There is no point in drawing lines which cannot be seen in the final drawing even though realism calls for them, and no purpose is served by drawing every hole in a large regular array of them even though there is a command in the software to do it for you. They will increase the time taken to plot, increase the storage size of the drawing and may make work on it slow due to the extra data which has to be searched.

The temptation for some to proliferate symbols has been mentioned elsewhere along with the importance for discipline in the use of layers.

If there is any kind of access control implemented involving special privileged passwords there are certain types who feel compelled to discover the passwords by fair means or foul.

Finally, the power of parameterised drawing has a fatal fascination for some. As is discussed in "Parameterised drawing and customising" on page 127 a good reliable parameterised drawing program takes a while to develop and the effort put into it can only be recovered by using it many times. The problem occurs when an enthusiast wastes time writing a program to do a drawing instead of actually doing the drawing. The remedy is to bring parameterised drawing under management control with properly approved and budgeted projects - the work being done by the one who is discovered to be the enthusiast!

THE CHORES

A CAD system generates a number of chores of its own which somebody has to be responsible for. Many of them can be done by junior or unqualified staff and the CAD/CAM manager needs to develop written procedures and the appropriate means of monitoring them.

A big source of chores is the pen plotter, particularly if conventional liquid ink drafting pens are used. Each pen must be rinsed out at the end of the day and put in the ultrasonic cleaner, which is an essential item of equipment. Before mounting in the plotter at the start of a run, each pen must be examined for wear using one of the special pen microscopes which are available. Wear on the tips is quite surprisingly rapid. A worn tip cannot be seen with the naked eye yet prevents the pen from writing properly. If a roll-fed plotter is in use then the ink level in the pens will have to be checked from time to time. It is usually only necessary to check the spare paper on the feed roll once or twice a day. At the end of the day the take-up spool will need removing and the plotted drawings cut off it.

Data security procedures will require regular mounting and dismounting of tapes or disks and initiating the software performing the back-ups. (See "Back-ups" on page 300.) The tapes or disks will have to be properly stored and documented. An accurate and detailed record of archived drawings will have to be maintained.

Plotting, printing, backing up and archiving require a stock of pens, paper, ink and tapes or disks to be maintained, preferably using some simple stock re-order level procedure to ensure that plotting is not held up because the paper has run out.

Some of the larger installations may want to record usage against project in which case an accounting run has to be done weekly or monthly as required. The printout has to be distributed appropriately. The list of account codes will have to be maintained as projects start and finish.

Finally, the whole business of hardware maintenance has to be managed, including the air cooling units if a minicomputer in its own room is used. The contracts will need annual review and renewal and down-time for preventative maintenance will have to be scheduled several times a year.

Software maintenance, if it is appropriate, will involve reporting bugs and getting fixes as they arise and scheduling down-time once or twice a year for new issues to be installed.

As can be seen, a CAD system involves a certain amount of work just to keep it running smoothly. A small system of three or four workstations would use up about a quarter of the time of a qualified person while a large system would require a full-time manager. In addition, about an hour a day of the time of a junior person would be needed for routine chores such as tape mounting and plotter attention.

EXERCISES

1. List standards in the order in which they should be established during the early phases of implementation.

2. Make out a weekly schedule of duties for a junior assistant in running the installation.

Chapter 32 Data security and contingency planning

In this chapter we will be concerned with the various ways in which CAD drawings can be accidentally destroyed and the procedures for preventing the loss. Any drawing costs far more than the medium holding it, whether paper, polyester film, disk or tape, on account of the labour invested in producing it. Losing a conventional drawing on film etc is not serious as it will most likely have been printed already so that a secondary master can be made from the print. A CAD drawing, on the other hand, cannot be reconstituted completely from a print on account of the precise but invisible numerical model lying behind it, of which the print is only a poor expression. There may even have been other data on invisible layers or in a parts listing which has not been printed at all. Consequently, a CAD drawing can only be stored as the computer file of binary numbers which it is.

WAYS OF LOSING DRAWINGS

Setting aside the loss of the entire installation by fire, the most devastating cause of data loss is a head crash on a disk since the entire contents of the disk are destroyed in an instant. Current high quality disk drives are designed for a mean time between failures of 100,000 hours or 11 years. Low cost Personal Computer disks may have a mean time between failures of as little as 20,000 hours or about 2½ years. Another cause of loss is accidental erasure by a user. On some systems, such an erasure can be recovered because the file is only marked as not available by the file management software until the space it occupies is needed. Removing the mark before the file is actually overwritten restores the file. The operating system, MS-DOS, used on Personal Computers works in this fashion. A third cause of loss, also due to human error, is misfiling or mistakes in identification. The exact location of a drawing in the computer file management system, on the tape or disk, or in the storage cupboards is mislaid either because a proper record of the location was not made or because the record that was made turns out to be erroneous. The problem is not much different to that found in paper-based systems and the remedies are the same: keep careful records and observe strict procedures.

All these ways of losing vital work can only be avoided by careful management procedures which will be detailed below.

BACK-UPS

The most important data security procedures are those for guarding against disk head crashes. The object is to put a copy of the data (back-up) on to a tape or disk which can be removed from the computer and stored in a safe place. Should the disk crash the data can be copied back on to the new disk and work resumed at the point when the copy was made. The work done between making the copy and the crash occurring is unfortunately lost forever. Clearly, the shorter the interval between making the copies the less is lost should a crash occur, but it is impractical continually to make back-ups every minute since each one takes time and uses up computing power.

The trend is to use larger and larger disks, and unless money is also spent on fast tape drives, making a copy of a complete disk can take hours and require several tapes. The problem is solved by omitting those disk files which never get updated or have not been updated recently. Once a file has been backed up, there is no need to take another copy until it has been changed. A good operating system will mark a file when it has been updated, thus allowing the back-up program to select it.

The back-up procedure therefore consists of first making a full back-up of everything, during which all the update marks are removed. Then, only those files which have changed subsequently are selected for back-up in what are termed "incremental back-ups". These are done as frequently as possible taking into account the amount of work which could be lost in a crash on the one hand and the resources of time etc required to do them on the other hand. As more and more files are updated, so each successive incremental back-up has to handle more files and therefore takes longer and uses up more tape or disk space. Eventually, a full back-up becomes worthwhile and the cycle repeats. To restore data after a crash the full back-up is replaced on the disk followed by the latest incremental back-up made before the crash. In addition to crash recovery, any accidental file deletion can be restored from either the full or the incremental back-up depending on when the file was last updated.

With these principles in mind, a detailed back-up procedure involving cycles of full and incremental back-ups can be devised. One scheme which has been used successfully is to do a full back-up once a week and an incremental twice every working day. If two incrementals can be put on one tape, a separate tape can be allocated for each day. With suitable software the daily tape can be mounted at the start of the day and an automatic back-up program started for the day which does incremental back-ups at appointed times. The tape is then dismounted at the end of the day.

A cause of data loss which is less likely but which one can still avoid is destruction of tapes by fire. There are two alternative precautions available. The cheapest is to keep two identical copies of each tape in two well separated buildings. The alternative precaution is to keep tapes in a large fireproof safe. The latter has the disadvantage of limiting the number of tapes that can be stored.

Accidentally deleted files can only be restored from full back-ups made before the deletion occurred, so if you wish to be able to restore many weeks after the deletion then several "back numbers" of full back-ups must be kept. This is usually done with a tape rotation scheme in which the latest back-up is overwritten on the tape currently holding the earliest back-up.

Using two copies in separate buildings also leads to a tape rotation scheme in which the tapes are rotated between buildings. If one building is the computer room, a convenient procedure is to write two tapes, keep one in the computer room and take the other to the other building exchanging it with the tape that was there. This leaves a copy of the latest back-up and the previous back-up in the computer room for restoring accidental deletions, and a second copy of the latest back-up in the other building to guard against fire. Variations on the scheme can be devised. For instance, six back numbers can be kept rotating in the computer room while two rotate between the computer room and the other building as shown in Figure 32.1. It is a good idea to write a simple program to keep track of the tapes and instruct the operator which to mount and which to take to the other building each time.

The idea of only backing up files which change can be taken a stage further by keeping the disk files containing the software separate from the drawing files and making special back-ups for them, since they only change when there is a new issue. Separate sets of tapes need to be kept and rotated every time a new issue has been installed and customised. Sometimes new issues of the operating system are installed at different times to the CAD software, in which case it needs its own set of back-up tapes. In both cases, two or three rotating tapes are sufficient. To allow the back-up program to make the right selection when it runs the disk will need to be properly divided up between operating system software, CAD software and user data. Included in user data for this purpose should be any software written in-house as this will change more frequently than the CAD software or operating system.

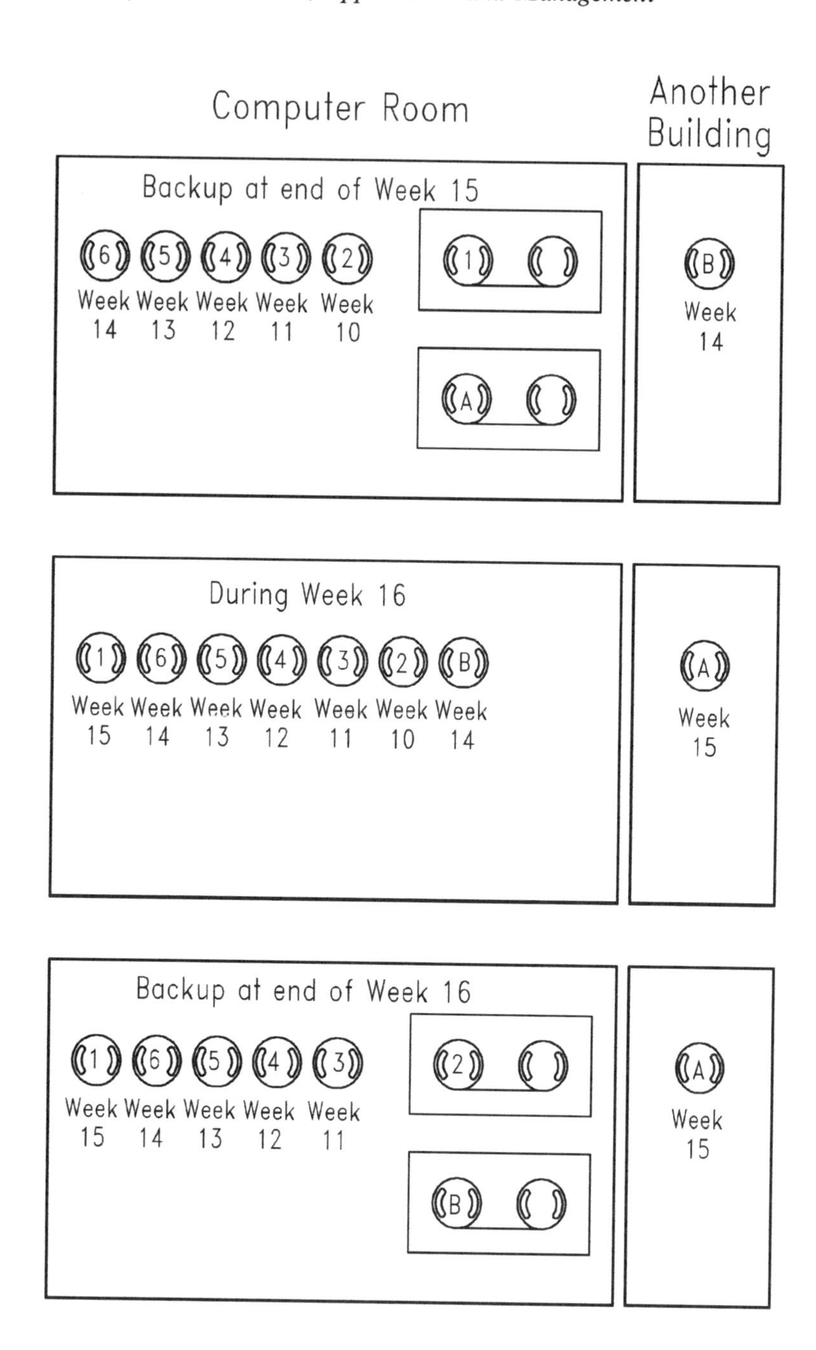

Figure 32.1 A tape rotation scheme

ARCHIVING

It is important to distinguish between archiving and backing up. The difference lies in the purpose for which they are done: in fact the way the data is organised on the removable medium is different. Back-ups are organised for restoring the entire contents of a working disk to its state at a particular time in the very recent past. Archive tapes or disks are organised for restoring a particular file to a working disk a long time after it was archived. Thus, archiving is geared towards long-term storage and easy retrieval of individual drawings. Another, very important, difference is that back-ups, being for the purpose of covering the possible destruction of files, must be duplicates of the files whereas archives, on the other hand, must not duplicate the files for the very important reason of avoiding duplicate masters. If an archived drawing is left on the disk, someone can modify it without changing the drawing reference. There are then two different versions of the same drawing either of which could be used at some time in the future. The archiving procedure is therefore the *transfer* of selected drawings to the removable medium: the drawings are first copied on to the medium and then deleted from the working disk provided the copy has been done without error. Thus, an archiving operation must consist of the following steps:

1. Copy the selected files on to the removable medium.
2. Verify that the copies are exact and free of errors.
3. Record the identifiers of the files copied including their sizes and any descriptive information which may be needed for retrieval.
4. Delete the files from the working disk.

The operation is best carried out by a program to ensure that none of the steps is omitted. It may well be desirable to make two copies for separate buildings as a precaution against loss by fire.

One aspect which may need some consideration is the use of "project" files containing data such as layer names, line drawing specifications or symbol definitions which drawings use but only store references to them within their own files. If a drawing is returned from archive after the project file has been changed the data it uses has changed. The project file data must be archived as well in such a way that it can be restored when the archived drawing is restored. This problem would be solved if the drawing was archived in a reliable neutral file format.

THE LONG-TERM STORAGE OF CAD DRAWINGS

Engineering drawings have a long life. There are engineering companies with drawings from the beginning of this century or earlier in their stores. Some products, particularly in civil engineering or capital goods, have a long life and it may be necessary to refer back to the design details any time during their life. Safety or patent litigation will require reference back to the procedure used in the design at any time during the life of a product.

One of the effects of CAD/CAM which few people realise is that the large archive of drawings on polyester film, vellum or cartridge paper which engineering companies keep will be joined by a tape or diskette library of drawings produced on the CAD/CAM system. The question is: "Will these drawings be as accessible in 50-100 years' time as the old paper drawings are?" Three conditions must be met to make this possible:

1. The contents of each tape or disk must be known in some detail. You cannot search through a tape or disk in the same way as you can through a filing cabinet.

2. There must be software at least capable of reading and displaying the tapes or disks in 50-100 years' time and preferably capable of plotting them out or converting them to the format in current use by the CAD system.

3. The tapes or disks must not have deteriorated in that period.

The easiest condition to meet is the first but there will be many CAD/CAM managers who will not have realised what is happening as they begin to accumulate a store of tapes or disks without keeping adequate records. Current CAD/CAM systems provide practically no facilities to support this vital activity.

The second condition raises further problems. CAD systems are developing rapidly as software is improved in response to market demands and hardware improvements. One of the consequences of development is a change in the format of the CAD drawing file. This is usually "upwards compatible". That is, the new version of the software will read the format of the previous version. However, it is not always the case that subsequent versions of the software will read the old format, still less when development over ten or more years has produced several successive formats. And all this is quite apart from the strong possibility that the company will have changed its CAD/CAM system over the period either from choice or because of a change of ownership, or that the original CAD/CAM system producer will have gone out of business and been disbanded. Drawings on paper can happily survive these changes because they are in a universal format. CAD drawings are currently in a private proprietary format. The problem can be overcome by converting all previous drawings to the new format when a change occurs, but as the archive gets bigger the work involved will become enormous. There are only two possible long-term solutions. One is the arrival of a reliable neutral format supported by international standardisation. The STEP format promises to fulfill this hope. The other is for all purchasers and users to insist on publication of the archival tape or disk format. This is not an unreasonable request as it still allows the supplier to use proprietary formats for the work files on the main disk.

The third condition also presents problems. Conventional drawings use well established materials such as drawing ink and paper and simple systems such as pigment adhering to sheet material. Tapes and disks use new materials such as ferric oxide powder embedded in a binder and the system is complex, consisting of digital codes in minute patterns of magnetisation. Paper and ink have been in use for hundreds of years but magnetic recording has hardly been in use 40 years. There is no way of avoiding the archival storage of CAD drawings on magnetic media. Paper plots only reproduce a fraction of the information contained in a CAD drawing. The safest way is to use the magnetic recording system which has been in successful use for the longest time - the half-inch 9-track magnetic tape - and follow the precautions recommended as a result of scientific research into the changes likely to occur in the medium over a long period. The following recommendations are taken from Reference (19), which is a comprehensive account of the care of magnetic tape and disk:

1. Tapes should be stored at 5-32°C and 20-80% relative humidity. The optimum is 18.3°C and 40% relative humidity. Note that the temperature range is wide but the humidity is critical. The particles of magnetic material and the backing medium of the tape are both stable but the magnetic particles are suspended in a flexible binder made of polyurethane which is hygroscopic and thus absorbs water from the atmosphere until its water content is in equilibrium with that of the atmosphere. The binder reacts chemically with the water so that its properties change. It becomes weaker so that it comes away from the backing, produces gummy products which adhere to the tape drive and it becomes more abrasive. Fortunately, the chemical reaction is reversible so that a tape can be rejuvenated by storing it in the correct environment again.

   Variations in temperature and humidity will change the tension in the tape reel. Insufficient tension results in cinching (see "The computer configuration" on page 59) and excessive tension results in adjacent turns of tape sticking together and even pulling the coating off.

2. Use the well-known recording oxide gamma ferric oxide as its ability to maintain its magnetisation over long periods is well established.

3. Use backcoated tapes. They provide better friction between layers and the signal level from them is more consistent.

4. Clean and test the tape (even if new) no more than six months before writing it. There should be no more than five write-skip errors, none of which should exist on the first 100 ft. The tape should have had at least four passes but no more than 50 before writing. Before writing the data, give the tape one full length pass at normal speed and do not rewind at high speed. The tension should be approximately 1.7-2.2N or 6-8ozf. Avoidance of tape debris and correct tape tension are important.

5. Avoid recording at 800 bpi recording density. (This density is becoming obsolete anyway.)

6. Ensure that the tape drive writing the tape is correctly adjusted, particularly with regard to tape tension. It should not introduce debris into the tape or wind it unevenly. Misaligned tape guidance components can create debris. The writing head should be correctly aligned.

7. Verify the recording with a read-only pass.

8. Fasten the leader with a sponge, grommet or vinyl strip.

9. Do not label the reel with a graphite pencil or water-soluble marker.

10. Put the tape in a fully enclosed, clean transparent container. It is important that the canister is properly closed.

11. Store the canister upright so that the tape is supported by its hub.

12. Ensure that no magnet can get closer than 3 inches to the tape. Other types of field such as high voltage, high frequencies, X-rays or nuclear radiation do not damage data.

13. Use what ever means is possible to protect the tape from fire. Carbon dioxide, Halon and water are permissible for fire extinguishing.

14. Other recommended practices are maintaining the storage temperature within 2.8°C and the humidity within 5% of the computer installation, and employing positive internal air pressure to reduce the invasion of dust.

15. Exercise each tape annually or semi-annually with a full forward and reverse pass at normal reading speed and tension. Do this on a special tape-cleaning machine if possible.

16. Visually inspect the tapes for irregular windings, creases, warped edges, discolouration etc.

17. Read samples of the tapes (or a set of tapes from the same manufacturing batch as the others specially written for test purposes) on an annual basis to check for read errors. Do this if possible in the archiving environment to avoid stresses caused by the change of temperature. A large number of temporary errors indicates deterioration and excessive debris in the tape. Tapes from the same manufacturing batch are likely to have deteriorated in the same way. At least clean the bad tapes but consider recopying them on to newer tapes.

18. Recopy all important tapes annually. Recopying all tapes at least once every ten years has been suggested.

19. When bringing tapes into the computer room, allow them to adjust to the temperature for about 24 hours.

20. Ensure all tape drive ferrous components are not magnetised in any way by using degaussing equipment (without any tapes in the vicinity!).

21. One or more of the following remedies may work if a tape cannot be read:

    - Use a back-up version of the tape if available.

    - Inspect the tape visually for irregular, folded, separated or protruding layers of tape and correct them by winding and rewinding twice completely at read/write speeds. Then keep the tape in the computer room for another 24 hours and wind and rewind several times again before attempting to read it.

    - Check the adjustment of the tape drive transport and look for wear.

    - Clean the tape drive transport. This is essential if the tape may have shed debris as a result of its physical defects.

    - Clean the tape.

- Read the tape on the drive which wrote it.
- Get a technician to try different mechanical and electronic adjustments on the tape drive.
- Try reading the tape in the reverse direction if the drive has this facility.

By way of summary, loss of data is due to just a few factors which the recommendations aim to overcome. These are:

- Debris and disintegration caused by softening of the binder by the water vapour in the atmosphere.
- Debris caused by physical maladjustment of the tape transport.
- Distortion of the tape backing while wound on the reel.
- Expansion and contraction of the tape backing due to temperature changes, creep and flow altering the tension in the reel.

It is interesting to note that there is little danger of losing the actual magnetisation of the oxide particles. The problems are all in the materials used to support the particles.

Reference (19) also describes procedures for monitoring tape tension throughout a reel and for testing binder integrity. The precautions require the use of special tape cleaners. These can be hired or the whole cleaning process can be performed by a service company on its own equipment. Another special machine used tests the tape and records errors to certify it.

ALTERNATIVES TO TAPE

Various other storage media will be invented as time goes on. A promising introduction was the optical disk. Unfortunately, it uses a very thin, highly reactive film protected by a plastic coating that is not completely impervious to water vapour. It takes time for the long-term stability of any newly invented medium to be established.

CONTINGENCY PLANNING

This chapter has been concerned with precautions against failure of the data storage medium. It is equally important to give some thought to complete loss of the entire installation due to major disasters such as fire. The entire work of the Design Department will come to a halt. Hopefully, the work done to date will not be lost due to the precautions discussed earlier. Should the entire installation be destroyed, something has to be found to replace it very quickly, and it is for this reason that many computer departments make plans to cover the event. One way is to make an arrangement with a similar installation for transferring the work there while the equipment is being replaced. There are even companies specialising in providing stand-by services. Whatever you decide to do it is worth planning the whole operation in detail (and then filing it away in the hope it will never be used!). The move to the alternative site, the software tapes that need to be taken etc all need some consideration.

This chapter has been a gloomy chapter of necessity but there is nothing like feeling that all possible disasters have been considered and an answer found before they happen. There is also a peculiar satisfaction in having survived a disk crash or other disaster unscathed!

EXERCISES

1. What bad consequences can arise from using a back-up procedure for archiving?

2. What is the most important characteristic of a tape store for reliable data storage?

3. Give the procedures to be followed with respect to tape reels for maximising the reliability of data storage.

Chapter 33 Programming practice

As it is quite likely that the person responsible for running or managing the CAD installation will do some programming, some guidance will be provided here on the best way of setting about the production of software.

WRITING FOR ONESELF AND WRITING FOR OTHERS

There is a big difference between writing a program to meet an immediate short-term requirement of your own and writing one to be used by several people over a period of time. It is what makes the difference between a professional programmer and an amateur. The differences lie in the following areas:

1. Handling of errors in input data

2. Maintainability

3. Documentation

User input dialogue

If you are writing for use by many other people you cannot assume anything about the data presented to the program even if you clearly specify what is required. This means that the programmer must thoroughly check the data before proceeding, inform the user clearly of any mistakes and give an opportunity to re-enter data which is wrong. It is also helpful to provide default values either based on commonly used values or the values entered the last time the program was run. Another feature which may be desirable is a "help" facility which the user can call upon to explain how to enter the data. The help text can be displayed automatically if an error is made. With all this, it is not unusual for the routine handling user input to account for more code than the rest of the program put together.

Maintainability

Such is the complexity of the typical program that it is quite possible for a programmer to return to something he has written six months earlier and be unable to understand it. Still less, therefore, can someone else understand it. Yet malfunctions or "bugs" will be discovered during the first six months of use which will have to be put right. Maintainability is that quality of programming which makes it possible for the program to be sufficiently well understood for it to be successfully modified by someone who did not write it in the first place. Maintainability also involves designing the internal structure of the program so that an alteration in one part does not produce unexpected results in another part. There are many ways in which this can occur and it is analogous to, for example, thickening a link in a mechanism and then finding that it clashes with another part as a result.

Documentation

Users must be provided with instructions on how to supply the data in written manuals, in the prompt messages or in a "help" facility. Futurc maintenance by a programmer is aided by notes and information on algorithms and data structures used, definitions of variables and flow charts.

THE STEPS IN PRODUCING A PROGRAM

As one might guess from the foregoing discussion, a professional approach to programming requires more than just writing code. The various stages can be outlined as follows:

1. Develop the specification
2. Define the variables
3. Define the structure
4. Write and test the code by module
5. Test the program overall
6. Document the program
7. Issue the program

Develop the specification

Because any program worth while producing is an extremely complex thing, specifying it properly is not easy. What one specifies is, in fact, a pattern of behaviour, a set of responses to various inputs. Some of it can be specified using mathematical formulae but not all of it, on account of the ability of computers to make decisions. A decision is, in effect, a discontinuity in a mathematical function. Some effort is needed therefore to ensure firstly that all the potential users have clearly defined what they want it to do. This is not always easy. Having got the users to define clearly what they want the programmer must then understand what they want it to do. There are thus two places where misunderstanding can occur: in the users' understanding of what they want and in the programmer's interpretation of their wishes. There is plenty of room for misunderstanding unless the programmer is very familiar with the problem the program is to solve.

A part of the specification which must not be forgotten is the way in which exceptional cases in the input data are to be handled. This includes defining what is exceptional data and then what the program is to do when it occurs.

Define the variables

A program achieves its end by manipulating a model of the real world. The first stage in designing a program is deciding how to represent the real world in terms of the types of variable available in the chosen programming language. Each quantity in the real-world problem will have to be assigned a variable and a physical unit of measure. Single numbers will be straightforward but arrays of numbers will need more thought. Logical variables representing "true" or "false" can cause problems if not carefully defined. In the author's experience, much trouble can be avoided by writing out a single sentence for each variable which defines its meaning in the real world.

Define the structure

Before embarking on writing code the internal complexity of the program should be got under control by establishing an organisational structure for it. The processing should be divided up into well defined tasks and the code for each task written in a clearly identifiable unit. It should be clear what data is calculated by each task and what data it uses for the purpose. The subroutine or procedure facilities available in the language should be employed to keep the tasks separate. The object is to be able to track down the source of an error when it occurs. The principle is the same as setting up a human organisation with precise responsibilities assigned to each department or team.

Write and test

Each task is now coded and tested on its own. Test data should be devised if possible which causes every possible path through the code to be taken. When every task has been done they can be tested together.

Document the program

The program should be documented. Most of the documentation will be in the comments written in the program. The purpose of a comment is to say something which is not apparent from the code, preferably the real-world interpretation of what is happening. The biggest source of problems is always the conditional statements. Comments stating the real-world condition under which each branch is taken are particularly valuable. Where the logic is complicated, a flow chart should be drawn. All variables should be declared even in languages like FORTRAN or BASIC which do not demand it. Against each variable declaration should be a comment elaborating its real-world meaning and unit of measure. (A useful technique in FORTRAN is to use a multi-line declaration in which each variable has a line with a comment.) A common source of bugs is not properly understanding the meaning of a variable.

Finally, the user needs a manual. Particularly important are precise and detailed definitions of the data to be supplied and the data which is generated by the program. It should be very clear how the output data has been derived from the input data. Error conditions should be documented in as much detail as possible either by extensive messages when they occur or by error numbers output by the program which are then explained in the manual with suggestions for avoiding the error. The only genuinely friendly program is the one which helps the user when he makes a mistake.

Issue the program

The final stage is issuing the program. If the program is supplied to a number of computers at different sites and is to be maintained and updated with bug fixes then it is important to know who has the program and, as updates are produced, which version. Issue consists of recording this information. The program itself should also indicate which version it is by a message on the screen or printout.

EXERCISE

What would you have to do to convert a program written for your own private use to make it available for general use?

A checklist of features

The list in this appendix summarises the account of CAD/CAM features given in the main part of the book. It is unlikely that any one system has all the facilities mentioned but the list could be used as the basis for developing a specification or set of criteria for selecting or evaluating systems.

HARDWARE

Computer

1. Type

   - Personal Computer
   - Self-contained workstation
   - Shared minicomputer
   - Mainframe

2. Networking

3. Removable storage medium

4. Maximum number of users before upgrade

5. Absolute maximum number of users

6. Maximum number of drawings in working storage

7. Absolute maximum number of drawings in working storage

Plotter

1. Media to be used

   - Paper
   - Vellum
   - Polyester film
   - Pre-printed

2. Sheet or roll or dual

3. Colour required?

4. Writing technique

   - Pen
   - Electrostatic
   - Ink jet
   - Laser

5. Maximum size of drawing

USER INTERFACE

Graphics screen

1. Size

2. Nominal resolution

3. Picture quality

4. Reduced reflections

5. Smooth movement of entities

6. Instant pan and zoom

7. Instant rotation in space

Pointing device

1. Tablet

2. Joystick

3. Roller ball

Menus

1. Tablet

2. Second screen

3. Graphics screen

4. Graphics screen and icons

5. Ease of navigation in menu tree

6. Programmed function keys

7. Help feature

8. Context-sensitive help feature

SYSTEMS SOFTWARE

1. File management

   - Multiple directories
   - File selection facilities
   - Automatic file version numbering
   - Degree of access control

2. Power and hardware failure resistance

   - Time to recover after power failure
   - Work lost due to power failure

3. Back-up management software

4. Archive management software

5. Usage monitoring

   - Level of detail in logs
   - Ceilings on storage use

6. Plot management

   - Plot queue management - plot deletion, rescheduling
   - Multiple plot queues
   - Pen scheduling
   - Software-generated line thickness
   - Automatic drawing frame merging

INTERACTIVE DESIGN FUNCTIONS

Fundamental software parameters

1. Basic resolution of numbers (significant figures)
2. 2D or 3D
3. Draughting facilities in a 3D system

Repertoire of entities

1. Point
2. Grid of points
3. Unlimited straight line
4. Single-segment line
5. Multiple-segment line
6. Circular arc
7. Conic section arc
8. High degree arc
9. Constraint (set of points and vectors constraining a high degree arc)
10. Composite line (mixed straight segments and arcs)
11. Closed outline
12. Subfigure
    - Copied
    - Referenced
13. Transformation
14. Alternative (local) axis system
15. Set (group)
16. Hierarchy or family tree (sets owned by sets)
17. Surface

- Cylindrical
- Spherical
- Ruled
- Sculptured
- Surface with arbitrary boundaries (face)
- Joined faces
- Enclosing a volume of space

18. Solid

- Faceted
- Precise
- B-Rep
- CSG

Non-geometric entities

1. Text

2. Dimension

- Length vertical, horizontal or any specified direction
- Multiple dimensions to a common datum
- Angle
- Radius
- Diameter
- Between concentric arcs
- Associative
- Circumferential
- Ordinate
- Axonometric
- Geometric tolerances

3. Arrow

4. Attribute

- Text string
- Number
- Multiple choice/binary

Viewing and presentation controls and facilities

1. Window definition
2. Arrangement of windows (screen)
3. Numbered layers
4. Named layers
5. Named lists of visible layers
6. Draughting view
7. Arrangement of draughting views
8. Multiple work areas
9. Line thickness, user selectable/definable
10. Line colour, user selectable/definable
11. Line styles, user selectable/definable
12. Filling (hatching) patterns, user selectable/definable
13. Perspective projections
14. Colour-shaded pictures

Definition methods

1. Coordinates

   - Digitised from tablet
   - Nearest grid point to cursor
   - Cartesian coordinates from keyboard, absolute
   - Cartesian coordinates from keyboard, relative
   - Mixed keyboard entry and digitised
   - Polar coordinates from keyboard, absolute
   - Polar coordinates from keyboard, relative
   - Intersection of two lines
   - Intersection of line and plane or surface
   - Limits or centre of an entity
   - Projection of a point on to an entity
   - Perpendicular to two lines

2. Straight lines

- Two points
- Direction, length and a point
- Subdividing an angle or the space between parallel lines
- Offset from another line
- Through a point and tangent to a curve
- Tangent to two curves
- Intersection of planes
- Axis of a cylinder
- Through a point and normal to a plane or surface
- Inferred from orthogonal views

3. Circular arcs

- Centre, radius/diameter and end points or angles
- End points and radius/diameter
- Three points on circumference
- Two tangents
- Offset from another arc

4. High order curves

- Constraining points and directions
- Offset from another arc
- Intersection of surfaces, planes

5. Surfaces

- Movement of a straight line/curve along two curves
- Rotation of a straight line/curve around an axis
- Generating curve in a generating plane moving along a spine
- Constraining points and directions

6. Faces

- Projection of lines on to a surface
- Intersection of surfaces

7. Solids

- Parameterised primitives
- Moving a closed outline along a straight line
- Moving a closed outline along a curve
- Rotating a line/curve round an axis
- Combining solids with Boolean operations

- Inferred from orthogonal views
- Definition in terms of engineering features (e.g. boss, hole, web)

Transformations

1. Translation, rotation, scaling, mirroring
2. Scaling along one axis only
3. Combination of transformations
4. Three-point definition of a multiple transformation
5. Move or copy
6. Storage of transformation for future use
7. Option to include, exclude or follow draughting rules for text
8. Option to apply it to constraining or end points only (line deformation)
9. Circular arrays
10. Rectangular arrays

Entity selection

1. The current entity
2. Enclosing rectangle
3. Enclosing polygon
4. Option to select those wholly within, wholly without or partially within the polygon
5. Lines broken at polygon
6. By layer
7. By attribute or non-geometric property
8. By entity type
9. By name
10. By set and family tree if allowed

11. Accumulation of a list of selected items
12. By a Boolean combination of selections
13. By draughting view

Symbol management

1. Multiple libraries
2. Exchangeable tablet menus
3. Connection nodes for schematic lines
4. Displayable attributes
5. Symbols definable with symbols

Line and curve editing

1. Trim to intersection
2. Break at intersection
3. Fillet at intersection
4. Chamfer at intersection
5. Reposition an end point
6. Extend line a desired amount or fraction
7. Boolean operations between overlapping polygons

Automatic dimensioning facilities

1. Numerical value calculated
2. Leader lines and arrows drawn and text centred between them
3. Intelligent placement of arrows inside or outside leaders
4. Intelligent stacking of overlapping dimensions
5. Complete automatic generation and placement of entire dimension
6. Generation of standard tolerance

7. Permanent association with element being dimensioned

Dimensioning style options

1. Selection of national or international standard
2. Numerical resolution
3. Units
   - Decimal inches or metric
   - Inches and fractions
   - Feet and inches
   - Dual units
   - Decimal degrees
   - Degrees, minutes and seconds
4. Tolerance style
   - Plus or minus
   - Upper and lower limit
   - Absolute
5. Forcing arrows outside or inside
6. Value in or above arrows
7. Leader line end position in relation to entity dimensioned
8. Text size and style

Text options

1. Height
2. Width
3. Spacing
4. Slant
5. Font
6. User-designed fonts
7. Multiple lines

8. Spacing between lines
9. Datum position relative to text
10. Fitting in a box
11. Automatic sequence numbers

Analysis

1. Area
2. Perimeter
3. Centroid position
4. Bending moments
5. Parameters of curves
6. Volume and weight
7. Surface area
8. Moments of inertia
9. Integrated formula calculator

DESIGN AUTOMATION

Parameterised drawing and user programming

1. Proprietary graphical language
   - Incremental compiler or interpreter
   - User prompting and input
   - Expression evaluation
   - Array variables
   - Conditional execution
   - Loops
   - Subroutines
   - Local variables in subroutines
2. Stable interface to a standard language
3. Report generator for attribute extraction

4. Variational geometry (graphically defined parameterisation)

Particular applications

1. Numerically controlled machine program generation
2. Mechanism design and simulation
3. Stress analysis of complex volumes
4. Stress analysis of frames
5. Thermal analysis
6. Injection mould analysis
7. Robot design and simulation
8. Sheet-metal design (development)
9. Sheet-material nesting
10. Reinforced concrete design
11. Structural steelwork design
12. Process plant design

References

1. Clarke, M J. "On the problems involved in turning a computer aided draughting package into a computer aided design system". *Proceedings of the Institution of Mechanical Engineers, Conference: Effective CADCAM '87*, p13.

2. Mavromihales, M and Weston, W. "Parametrically drawn shafts for an electric motor manufacturer". *Proceedings of the Institution of Mechanical Engineers, Fourth European Conference: Effective CADCAM*, p21.

3. Wells, C S. "The design supervisor - a changing role with CAD?". *Proceedings of the Institution of Mechanical Engineers, Conference: Effective CADCAM '87*, p27.

4. Smith, P J. "CADCAM data exchange - what it is and how to make it happen". *Proceedings of the Institution of Mechanical Engineers, Conference: Effective CADCAM '87.*

5. Blackburn, D J. "CADCAM communications - a practical approach". *Proceedings of the Institution of Mechanical Engineers, Conference: Effective CADCAM '87.*

6. *Initial Graphics Exchange Specification (IGES), Version 4.0*, available from CADCAM Data Exchange Technical Centre, 177 Woodhouse Lane, Leeds, LS2 3AR.

7. Adobe Systems Incorporated. *PostScript Language Tutorial and Cookbook*, Addison-Wesley, 1985.

8. Adobe Systems Incorporated. *PostScript Language Reference Manual*, Addison-Wesley, 1985.

9. Carlbom, I and Paciorek, J. "Planar geometric projections and viewing transformations". *ACM Computing Surveys*, December 1976.

10. Freeman, P F and Wesolowski, J P. "Turbocharger compressor wheels from concept to customer samples - an integrated design system". *Proceedings of the Institution of Mechanical Engineers, Fourth European Conference: Effective CADCAM*

11. Chang, C and Melkanoff, M A. *NC Machine Programming and Software Design*, Prentice-Hall, 1989.

12. ANSI X3.37-1987. *Programming Language APT*, American National Standards Institute.

13. ISO 3592-1978. *Numerical Control of Machine Tools - NC Processor Output - Logical Structure (and Major Words).*

14. ISO 4343-1978. *Numerical Control of Machine Tools - NC Processor Output - Minor Elements of 2000-type Records (Post Processor Commands).*

15. ISO 6983/1-1982. *Numerical Control of Machines - Program Format and Definition of Address Words - Part 1: Data Format for Positioning, Line Motion and Contouring Control Systems.*

16. RS 274-D. *Interchangeable Variable Block Data Format for Positioning, Contouring, and Contouring/Positioning Numerically Controlled Machines*, Electronic Industries Association, 1979.

17. RS 494. *32 Bit Binary CL Exchange (BCL) Input Format for Numerically Controlled Machines*, Electronic Industries Association, 1983.

18. Kief, Hans B. *Flexible Automation 87/88*, Baeker Publishing Co (UK).

19. Geller, S B. *Care and Handling of Computer Magnetic Storage Media*, National Bureau of Standards, 1983.

20. Mullineux, Glen. *CAD: Computational Concepts and Methods*, Kogan Page, 1986.

21. Seyer, Martin D. *RS-232 Made Easy: Connecting Computers, Printers, Terminals and Modems*, Prentice-Hall, 1984.

22. Honig, David A. *Desk-Top Communications: IBM PC, PS/2 & Compatibles*, Wiley, 1990.

23. Scheifler, Robert W. *X Window System: C Library and Protocol Reference*, Digital Press, 1988.

24. Hall, Frank E. "X: A window system standard for distributed computing environments". *Hewlett Packard Journal*, October 1988.

25. Newman, W H and Van Dam, A. "Recent efforts towards graphics standardisation". *ACM Computing Surveys*, December 1976.

26. Bergeron, R D, Bono, P R, and Foley, J D. "Graphics programming using the CORE system". *ACM Computing Surveys*, December 1976.

27. Salmon, R and Slater M. *Computer Graphics: Systems & Concepts*, Addison-Wesley, 1987.

28. Stark, J. *Managing CAD/CAM: Implementation, Organisation and Integration*, McGraw-Hill, 1988.

29. Weatherall, A. *Computer Integrated Manufacturing: From Fundamentals to Implementation*, Butterworths, 1988.

30. Sayle, M and J, "Far East of Eton". *The Spectator*, 12th May 1990.

31. Brooks, I S and Wells, C S. "Role conflict in design supervision". *IEEE Transactions on Engineering Management*, Special Issue Part 2, November 1989.

32. Swift, K G and Firth, P A. "Knowledge based expert systems in design for automatic handling". *Assembly Automation, 5th International Conference*, 1984.

33. Hernani, J T and Scarr, A J. "An expert system approach to the choice of design rules for automated assembly". *Assembly Automation, 8th International Conference*, 1987.

34. Faux, I and Pratt, M. *Computational Geometry for Design and Manufacture*, Horwood, 1979.

35. Woodwark, J. *Computing Shape*, Butterworths, 1986.

36. Lindgren, Richard K. "Politics and system justification". *Computerworld*, Vol 17, No. 2, 1983.

37. Farin, Gerald. *Curves and Surfaces for Computer Aided Design*, Academic Press, 1988.

38. Davies, P and Anderson, J. "Knowledge-based engineering systems in practice". *Proceedings of the Institution of Mechanical Engineers, European Conference: Effective CADCAM '91*.

39. Krstin, L. "A user's experience of three-dimensional computer modelling for process plants". *Proceedings of the Institution of Mechanical Engineers, European Conference: Effective CADCAM '91.*

40. European Council. *Council Directive, 29th May 1990, on the minimum safety and health requirements for work with display screen equipment.* Reference 90/270/EEC.

Index

A

B

C

D

E

F

G

H

I

J

K

L

M

N

O

P

R

S

3